ADHESIVES TECHNOLOGY COMPENDIUM 2021

adhesion ADHESIVES & SEALANTS
Industrieverband Klebstoffe e. V.
(German Adhesives Association)

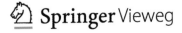

Springer Vieweg

Editor:

adhesion ADHESIVES & SEALANTS
Abraham-Lincoln-Straße 46
D-65189 Wiesbaden
Phone +49 (0) 6 11-78 78-2 83
www.adhaesion.com
Email: adhaesion@springer.com

Supported by:
Industrieverband Klebstoffe e. V.
German Adhesives Association
Völklinger Straße 4 (RWI-Haus)
D-40219 Düsseldorf
Phone +49 (0) 2 11-6 79 31 10
Fax +49 (0) 2 11-6 79 31 33

Publishing house:

Springer Vieweg | Springer Fachmedien Wiesbaden GmbH
Abraham-Lincoln-Straße 46
D-65189 Wiesbaden
www.springer-vieweg.de

| Austria (A) | Swiss (CH) | Germany (D) | Netherland (NL) |

Layout: satzwerk mediengestaltung · D-63303 Dreieich

ISBN 978-3-658-35377-3

Nominal sum: € 25.90

Dear Reader,

Today, almost every sector of industry and business relies on the use of adhesive bonding, which is an innovative and reliable joining method. It is essential when it comes to joining different materials while preserving their properties and offering long-term stability. The opportunities for using new, reliable construction methods can only be exploited in conjunction with innovative adhesive systems. In addition to the joining process itself, other functionality can also be integrated into bonded components, for instance by balancing the differing dynamics of joined parts, conducting electricity or heat, providing corrosion protection or vibration absorption or sealing against liquids and gases. More than any other joining technology, adhesive bonding allows complex designs to be created, because it offers the best possible combination of technological sophistication, cost-effectiveness and a low environmental impact. Adhesive bonding is without a doubt the key technology of the 21st century.

The German adhesives industry is particularly important in this respect. It is a technology leader on both European and global markets. The demand from other countries for adhesives and sealants that are made in Germany remains high. The German adhesives industry exports almost 45 % of its products every year. In addition, the foreign subsidiaries of German adhesives manufacturers generate sales of more than € 8 billion, which gives the German adhesives industry an almost 20 % share of the global market with adhesives and sealants designed in Germany.

Alongside its leading technological position, the German adhesives industry is also highly important in economic terms, which is something that the general public is almost entirely unaware of. Every year more than 1.5 million tonnes of adhesives and sealants and more than 1 billion square metres of adhesive films and tapes are manufactured in Germany, which results in total industry sales of almost € 4 billion. The use of these adhesive systems generates potential added value of more than € 400 billion. This huge amount corresponds to around 50 % of the contribution made by manufacturing industry and the construction sector to the German gross domestic product (GDP). In other words, around 50 % of the goods produced in Germany have a connection with adhesives and sealants.

The German Adhesives Association (IVK) represents the technical and economic interests of 154 manufacturers of adhesives, sealants, raw materials and adhesive tapes, key system partners and scientific institutes. It is the world's largest and, in terms of its comprehensive portfolio of services, the world's leading association for adhesive bonding technology.

In this compendium the German Adhesives Association, in cooperation with its sister associations – the Association of the Adhesives Industry in Switzerland (FKS), the Association of the Austrian Chemical Industry (FCIO) and the Association of the Dutch Adhesives and Sealants Industry (VLK) – gives an insight into the world of the adhesives industry.

One of the main tasks of these associations is to provide regular information about adhesive bonding as a key technology, as well as about manufacturers of innovative adhesive systems and the activities of the industry's organisations. This compendium contains important facts about the adhesives industry and its associations, as well as describing the extensive product and service profiles of adhesives manufacturers, key system partners and scientific institutes.

Together with the editorial team of "adhäsion KLEBEN & DICHTEN", we are pleased to present the 7th edition of our Adhesive Technology Compendium. The compendium is updated every year and published alternately in German and in English.

Dr. Boris Tasche
Chairman of the Board of Directors of
Industrieverband Klebstoffe e.V.

Dr. Vera Haye
Managing Director of
Industrieverband Klebstoffe e.V.

German
Adhesives
Association

Industrieverband Klebstoffe e.V.

adhesion ADHESIVE®
SEALANTS

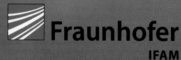

Fraunhofer

IFAM

EMICODE®

GEV

*Industrieverband
Klebstoffe e.V.*

List of Advertisers

Cover: Application of one-component epoxy resin adhesives for connector sealing © Panacol

Measuring and Testing

Process-Oriented Rheology of Packaging Adhesives

Rheological methods are widely used in the quality control of adhesives. In order to better reproduce the real process, however, interactions between adhesive and substrate must be considered. By selecting a suitable setup as well as multi-task tests, an optimized evaluation of adhesives is possible.

Modern rheometers allow us to characterize the complete curing process of the adhesive by means of oscillatory measurements [2]. This includes the characterization of the material from its initial liquid state (practically a Newtonian fluid) to its final cured state (in many cases practically purely a Hookean solid). Oscillatory measurements directly provide mechanical properties of the adhesive and can be combined with precise temperature and even humidity controls. In addition, modern rheometers allow accurate measurement and control in normal forces and gap widths. For these reasons, the use of rheometers in quality control has spread massively during the last decade in the adhesive industry. By using a small amount of material, the user can save a considerable amount of time and energy when characterizing the performance of the material. Depending on the type of adhesive, chemical reactions or other physical effects can influence the setting of the adhesive. Adhesives shrink, for instance, due to the release of water, or expand, due to the formation of CO_2. These phenomena are influenced by the type of substrate and, in turn, affect both the mechanical properties of the adhesive compound and the curing kinetics. Therefore, when adhesive materials come into play, considering the correct substrate may be essential for proper material characterization.

In the case of dispersion adhesives in the packaging industry, the cardboard absorbs the water and thus decisively influences the curing kinetics of the adhesive. It thus seems logical to consider the substrate in the rheological measurements if we want to closely simulate the real process in the performance of the dispersion adhesive.

This article provides an overview of how a rheometer with some practical adaptations can be used to measure and evaluate dispersion adhesives more effectively and more closely to the real process. A multi-task test is presented here with the objective of obtaining as much information as possible about the adhesive and its interaction with the substrate by means of a single measurement.

Adjusting the measuring geometries

If the surface of the material to be bonded influences the curing of the adhesive, both the measuring geometry and the bottom plate of the rheometer should first be slightly adapted. For the particular case of the packaging industry, this is especially relevant. If we use a standard measuring geometry and a standard bottom plate, the dispersion agent is trapped between the metal surfaces and nothing happens except curing at the edges of the geometry. Similar studies on including the surface in the rheological measurement are found in literature in the field of wood adhesives [2]. These techniques were further developed and discussed in more recent papers [3, 4, 5]. Figure 1 shows a setup for adapting the rheometer to include the appropriate surface for the real curing kinetics.

A layer made of the desired material is carefully glued to the exchangeable plate (E). The exchangeable plate is fixed then to the bottom plate of the rheometer using the clamping frame. The desired material is also glued to the disposable measuring geometry. Thereafter, the disposable measuring geometry containing the material layer is coupled to the upper shaft (A) by means of a fixing screw.

The surface material to be studied (substrate) should be glued to the support plate with an adhesive of high stiffness. Low-viscosity cyanoacrylate adhesives and 2-component epoxy resins have proven effective. The advantage of the setup is that any kind of surfaces may be glued to the bottom plate surface. Cardboard was chosen for studying the rheological behavior of adhesive dispersions in the present study.

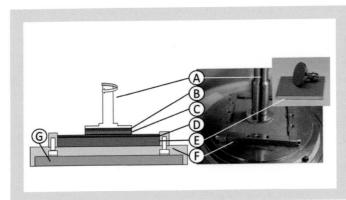

Figure 1: Adaptation of the rheometer to include top and bottom surfaces of the desired material: (A) upper shaft for disposable measuring geometries, (B, D) glued-on cardboard cover layer, (C) adhesive to be studied, (E) disposable plate, (F) inset including the clamping frame, (G) Peltier heating element.

© HNEE

Considering potential interactions

It is convenient to make a previous study on potential interactions between the adhesive and the adapted measuring system (containing both upper measuring geometry and lower plate). This is important to verify that undesired interactions do not affect subsequent measurements. Figure 2 illustrates the possible interactions between the different materials of the adapted system. For illustrative purposes, the system consists of the disposable measuring geometry material (A), the adhesive to be studied (B), the cardboard (C) and the adhesive used to fix the cardboard to the bottom plate (D).

An undesired interaction between the disposable measuring geometry and the adhesive to be studied can occur during long-term measurements if the disposable measuring geometry is made of Aluminum and the adhesive is very acidic. To cite an example, the acetic acid contained in polyvinyl acetate (PVAc), a common dispersion adhesive in the wood and paper industry, reacts with aluminum to form aluminum acetate and hydrogen. This reaction may distort the experimental results. Note that in general it is desired to characterize the interaction between adhesive and substrate (and not between measuring geometry and adhesive). Another example of undesired interaction is the one generated between the adhesive used to fix the cardboard to the plate and the cardboard itself. (Figure 2c and Figure 2d). In this case it is necessary to look not only for a fixing adhesive that provides high stiffness, but also one that does not penetrate into the substrate. Figure 2e shows an illustrative example of partial penetration of the adhesive used for fixing the cardboard into the cardboard itself. This penetration may reduce the absorption of the dispersion agent. The diffusion of the dispersants into the substrate material is a clear example of further interaction. This is actually the main interaction we want to consider to simulate the real process. To cite a second practical example, if we work with one-component polyurethane prepolymers it would be convenient that water absorption on the surface of the substrate is possible since it is necessary in the curing of the material [3]. Another possible interaction to consider is the one generated between the adhesive to be studied and the card-

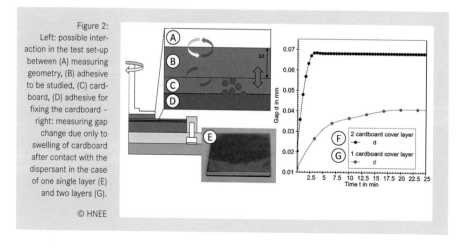

Figure 2:
Left: possible interaction in the test set-up between (A) measuring geometry, (B) adhesive to be studied, (C) cardboard, (D) adhesive for fixing the cardboard – right: measuring gap change due only to swelling of cardboard after contact with the dispersant in the case of one single layer (E) and two layers (G).

© HNEE

board fixed in the upper measuring geometry (the cardboard top layer). If the dispersant pene-
trates into the cardboard top layer, the material swells and changes the measurement gap. The
effect of this interaction can be thus studied by means of a simple experiment performed with
the rheometer. Figure 2f and Figure 2g compare an experiment performed with top and bottom
cardboard (black curve) with an experiment performed with only the bottom cardboard (red
curve). In the experiment, the upper measuring geometry was set to a gap of 0.01 mm after
wetting the lower surface with the dispersant. The normal force was set to 0 N and the variation
of the gap during the test was monitored. The results clearly show how each of the cardboard
layers swells around 30 – 40 µm. The change in the gap due to the swelling of the cardboard
layers can be considered for corrections in further experiments.

Gap control is actually a very important factor to consider during the curing of adhesives. Figure 3
illustrates how the gap filled with adhesive may change depending on whether we predefine a
constant gap, or whether we compensate the gap by controlling the normal force. In the case of
adhesives that expand (e.g. due to gas formation), foaming occurs in either scenario. In this
case, keeping the gap constant may bring the experiment closer to the real bonding process.
Figure 3a shows a cross-section of such a foamed edge generated by keeping the gap constant
during the experiment, and Figure 3b displays another case in which the gas pressure generated
during curing forced a large amount of material out of the measuring gap. In the case of shrinking
adhesives, compensating this shrinkage by normal force is advantageous. Here edge constrictions can be avoided (Figure 3c) and a quantitative measurement is possible.

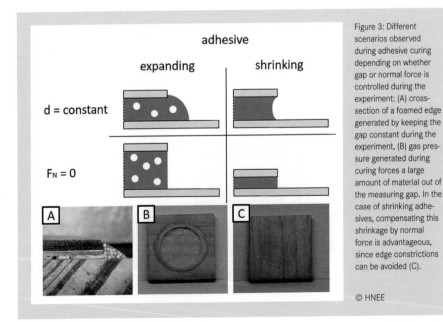

Figure 3: Different scenarios observed during adhesive curing depending on whether gap or normal force is controlled during the experiment: (A) cross-section of a foamed edge generated by keeping the gap constant during the experiment, (B) gas pressure generated during curing forces a large amount of material out of the measuring gap. In the case of shrinking adhesives, compensating this shrinkage by normal force is advantageous, since edge constrictions can be avoided (C).

© HNEE

Multi-task test

The multi-task test is intended to obtain as much information as possible regarding cohesive and adhesive properties of the adhesive-substrate system in a single measurement [6]. As an example, an application example of a multi-task test consisting of a time sweep, a frequency sweep and a final tack test will be shown here.

All three measurements are performed in series using the same specimen. The ultimate purpose is to obtain relevant material information efficiently so that the user can save as much time, effort and material as possible in characterizing the performance of his material.

The aim of the time sweep is to characterize the curing kinetics of the adhesive when it comes into contact with the substrate. Besides, valuable information about the convective or diffusive transport mechanisms through the substrate, i.e. penetration of the adhesive or dispersant into the cardboard can be obtained.

The frequency sweep is ideal for characterizing the cohesive properties of the material over a wide range of time scale. This information is well suited for characterizing the bonding and debonding behavior of the adhesive-substrate system. The long-term behavior (low frequencies) characterizes bond formation and the short-term behavior (high frequencies) characterizes debonding. The latter describes, for example, the peel adhesion of pressure sensitive adhesives and the former refers to properties such as shear strength and tack [7, 8, 9].

The last test, after the adhesive has partially cured, is based on applying a constant vertical speed and measuring the resulting force. In this test, the user can obtain some additional information about the adhesion properties.

To study the effectiveness of the multi-task test applied with the rheometer adaptation, two adhesive samples based on aqueous polyvinyl alcohol-stabilized vinyl acetate-ethylene copolymer dispersions were measured (sample 1 and sample 2). According to feedback from the market, sample 1 shows higher adhesive and cohesive properties than sample 2 in the real process. In other words, customers using sample 2 are most satisfied with the performance of the material when using it as an adhesive in the packaging industry.

Time sweep

In order to investigate the kinetics of the curing reaction, the samples were measured at a constant frequency of 1 Hz and a constant strain of 1 %. This strain remains within the linear viscoelastic range (LVE) of the sample. This is of critical importance in order to perform a non-destructive test. Therefore, an amplitude sweep was performed as a preliminary test to define a proper strain for the following experiments (not shown in this report). The curing reaction was performed at a constant temperature of 20 °C. A normal force of 0 N was set during the experiment. The change in the gap was used as an indicator of the expansion or compression of the sample (and cardboard).

The experimental results are shown in Figure 4a (storage and loss modulus), and Figure 4b (network formation rate and gap). As expected, G' and G'' show a continuous increase during the experiment due to the curing reaction of the adhesives. The initial values of G' and G'' are similar for both samples. However, after 30 min, the storage modulus G' has developed differently in the two samples. Sample 1 reaches a G' value of ~0.25 MPa while sample 2 reaches a value of ~0.08 MPa. After 30 minutes, both adhesives in contact with cardboard have reached the asymptotic zone with respect to the storage modulus. This means that although the samples have not fully reached a constant plateau in G', the three-dimensional network is already consolidated. No major changes in the shear modulus are expected after 30 min and therefore this time was used as a limit before moving on to the second test and evaluating the cohesive properties of the adhesive-substrate system. It should be noted that this time must be adjusted depending on the application of the adhesive.

The incorporation of the aqueous phase into the board leads to swelling of the substrate, which is also shown in the data. The gap as plotted in Figure 4b shows an increase in the first 5 min, suggesting expansion of the cardboard. For this particular case, shrinkage of the adhesive would be expected as a consequence of the curing reaction. The increase in the gap therefore suggests that the swelling of the cardboard is dominant over the shrinkage of the adhesive. The decrease on the gap after this first swelling phase may be the consequence of a slight shrinkage of the adhesive film due to loss of the aqueous disperse phase or to a decrease in swelling of the cardboard.

To better analyze the temporal evolution of each adhesive, it is convenient to plot the numerical differentiation of the storage modulus with respect to the time (network formation rate) (Figure 4b). Changes in the curves plotting the network formation rate correlate with specific internal morphological changes during the reaction [4]. In the first 4–5 min, both adhesives show an

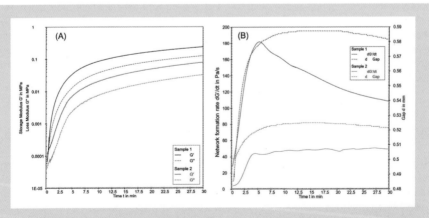

Figure 4: Time sweep for sample 1 (black) and sample 2 (red): storage and loss modulus (A), network formation rate and gap as a function of time (B). © Anton Paar Germany GmbH

increasing network formation rate. Sample 1, however, shows a considerable higher value at the beginning. This indicates that at the beginning of the reaction the curing kinetics is faster in sample 1. After about 5 min, sample 1 shows a monotonous decrease in the network formation rate while sample 2 remains at the same but much lower level.

The network formation rate is particularly dependent on the concentration and diffusion speed of the adhesive or dispersant. Different formulations should therefore have an influence on this parameter, which in turn can strongly influence the adhesive and cohesive properties of the cured material. This influence on the kinetics, including diffusion of material into the substrate, can be characterized on the first phase of the multi-task test. In our specific example, both samples show clear different behavior once they come in contact with the cardboard. This behavior in the initial phase will have a significant impact on subsequent tests.

Frequency sweep

After the material is partially cured on the cardboard, the frequency sweep appears to be an ideal test for characterizing the viscoelastic properties of the bulk material within a broad range of time scale. This is especially interesting, because this behavior on the short- or long-term time scale can be related to the performance of the material in typical tests used to characterize the performance of adhesives. Some literature, for instance, relates the rheological information at low frequencies to the classical tests used in the industry to characterize the shear strength of the material [8, 9, 10]. In general, the higher the storage modulus at low frequencies (f ~ 0.1 Hz), the better the shear strength of the material. Similarly, the rheological information at high frequencies (f ~ 100 Hz) is related to the peel adhesion of the material. The peel adhesion (debonding strength) is composed of the cohesive strength represented by the storage modulus G' and the dissipation energy indicated by the loss modulus G'. In general, therefore, the higher the G' and G'' values at frequencies near 100 Hz, the better the performance of the material in the peel adhesion test.

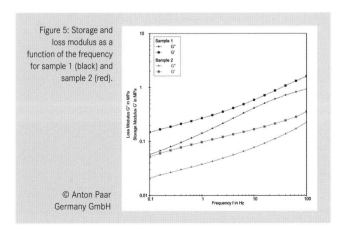

Figure 5: Storage and loss modulus as a function of the frequency for sample 1 (black) and sample 2 (red).

© Anton Paar Germany GmbH

In the multi-task test, the frequency sweeps were performed in a range between 100 Hz and 0.1 Hz at a temperature of 20 °C. A constant strain within the LVE range was set for the experiments. Figure 5 depict the storage and loss modulus for samples 1 and 2 as a function of frequency. For a frequency close to 0.1 Hz, the storage modulus of sample 1 (0.15 MPa) is higher than that of sample 2 (0.06 MPa). This indicates a higher shear strength of sample 1 compared to sample 2. From Figure 5, it can be also discerned that sample 1 has the highest G' and G'' in the high frequency range. According to this, sample 1 would tend to have the better performance in terms of peel adhesion.

Tack test

In the last step of the multi-task test, the upper measurement geometry is lifted with a constant speed of 500 µm/s. The measurements were performed at a temperature of 20 °C. During the experiment, the normal force was measured as a function of gap. The measuring point duration was defined by means of a logarithmic ramp from 0.02 to 0.2 s. This is useful to increase the resolution in the small gap region, where adhesion may play a more predominant role.

The maximum absolute force and the area under the normal force curve were used as evaluation criteria. The maximum absolute force is a direct measure of cohesion (tensile strength) and the area in the normal force-displacement diagram is a measure of the required separation energy, i.e. adhesion at the interfaces (tack) [1].

Figure 6 depicts the results of the tack test for both samples. The colored areas represent the separation energy calculated as the area below the normal force curve. The required separation energy is 32.4 mJ for sample 1 (black area) and 19.4 mJ for sample 2 (red area). The maximum absolute normal force of sample 1 is 17 N and 18 N for sample 2. For illustrative purposes, a measurement without a modified measuring geometry (i.e. measuring geometry and bottom

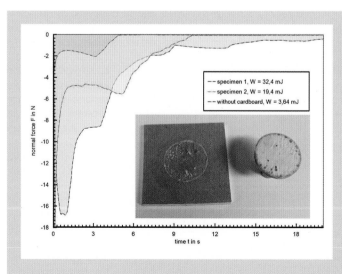

Figure 6: Tack test for sample 1 (black), sample 2 (red) and sample 1 using a metal measuring geometry and a metal bottom plate.

--- specimen 1, W = 32,4 mJ
--- specimen 2, W = 19,4 mJ
-- without cardboard, W = 3,64 mJ

© HNEE

plate without cardboard) is also shown in the Figure 6 (blue area). For that case, both measured separated energy and maximum normal force show considerable smaller values, 3.64 mJ and 3.6 N, respectively.

According to the experimental results, sample 1 has the best adhesion properties (larger area under the curve). The results obtained from the tack test are in good agreement with previous results. Sample 1 shows the highest storage modulus G' after 30 min (Figure 4a) and the highest value in G' at low frequencies (Figure 4b). In addition, sample 1 shows higher viscous properties represented by G''/G' at frequencies above 1 Hz (Figure 4b). This result may indicate that higher energy is thus required to release the adhesive 1 from the substrate (Figure 5). It should be noted that the higher the viscous contribution, the higher the ability of the adhesive to 'penetrate' into the substrate and therefore the higher the possibility to form a good bonding with the substrate (better tack). These results are consistent with the results obtained from the tack test. This interpretation is only possible if we consider the entire adhesive-substrate system in our experiments.

Conclusions

It could be shown that a small adaptation of the rheometer can be useful to better characterize the cohesive and adhesive properties of the complete system consisting of adhesive and substrate. The ultimate goal is to simulate the real process conditions as close as possible. A multi-task test is also presented in order to obtain as much information as possible about our system with a single measurement. For this purpose, three individual experiments are connected in series and run on the same specimen on the rheometer. The conclusions obtained from the multi-task test are in good agreement with the feedback obtained from the market.

Thanks

The authors would like to thank the company Eukalin Spezial-Klebstoff Fabrik for providing the samples and their experience with regard to the performance of the material.

References

[1] Mezger, T.G.: Das Rheologie Handbuch. Für Anwender von Rotations- und Oszillations-Rheometern. Vincentz Network, Hannover (2012)

[2] Witt, M.: Novel plate rheometer configuration allows monitoring real-time wood adhesive curing behavior. Journal of Adhesion Science and Technology 18 (2004) [8] 893- 904

[3] Winkler, C., Schwarz, U., Senge, B.: Materialwissenschaftliche Analyse des Abbindeverhaltens von Holzklebstoffen am Beispiel von 1K PUR. Chemie Ingenieur Technik 92 (2020) [6] 759 - 768

[4] Stapf, G., Aicher, S., Zisi, N.: Curing Behaviour of Structural Wood Adhesives - Parallel Plate Rheometer Results. The International Conference on Wood Adhesives, Toronto (2013)

[5] Schmidt, M., Knorz, M., Wilmes, B.: A novel method for monitoring real-time curing behaviour. Wood Science and Technology 44 (2010) [4], 407-420

[6] Agudo, J.R., Winkler, C., Schäffler, M.: Ein rheologischer Multi-Task Versuch für die Charakterisierung in der Klebstoff-Anwendungstechnik. Application Report C92IA060DE-A (2020)

[7] Chang E.P.: Viscoelastic windows of Pressure-Sensitive Adhesives. Adhesion 34 (1991) 89 - 200

[8] Chu, S.G.: Dynamic Mechanical Properties of Pressure Sensitive Adhesives. In: Lee, L. H. (Hrsg.): Adhesive Bonding. Plenum Publishing, New York (1991) 97 - 138

[9] Sun, S., Li, M., Liu, A.: A review on mechanical properties of pressure sensitive adhesives. International Journal of Adhesion and Adhesives 41 (2013) 98 - 106

[10] Winkler, C., Agudo, J.R., Schäffler, M.: Rheologische Charakterisierung der Vernetzung feuchtereaktiver Polyurethan-Prepolymere. Application Report C92IA053DE-A (2020)

The Authors

Christoph Winkler, M.Sc.
is a Research Associate in the Department of Wood Engineering at the University for Sustainable Development Eberswalde (HNEE).

Dr.-Ing. José Alberto Rodriguez Agudo
works as Application Scientist Rheometry and DMA at Anton Paar Germany GmbH in Ostfildern.

Prof. Dr.-Ing. Ulrich Schwarz
is Dean of the Department of Wood Engineering at the University for Sustainable Development Eberswalde (HNEE).

Michael Schäffler
- corresponding author -
(michael.schaeffler@anton-paar.com)
is Market Development Manager at Anton Paar Germany GmbH in Ostfildern.

COMPANY PROFILES

Adhesive Producer
Raw Material Supplier

A D H E S I V E   T E C H N O L O G I E S

Adtracon GmbH
Hofstraße 64
D-40723 Hilden
Germany
Phone +49 (0) 21 03-253 170
Fax +49 (0) 21 03-253 1719
Email: info@adtracon.de
www.adtracon.de

Member of IVK

Company

Year of formation
2002

Size of workforce
15

Ownership structure
Dr. Roland Heider, ISB, KfW

Sales channels
Direct and distributors

Contact partners
Management:
Dr. Roland Heider

Further information
Adtracon specializes in the development, production and marketing of reactive hot melt adhesives. We also offer technical and laboratory services and consultancy services.

Range of Products

Types of adhesives
Reactive hot melt adhesives

For applications in the field of
Bookbinding/graphic design
Woodworking/furniture industry
Automotive industry, aviation industry
Textile/filter/shoe and leather industry

ALBERDINGK BOLEY

Alberdingk Boley GmbH
Düsseldorfer Straße 53
47829 Krefeld, Germany
Phone +49 (0) 21 51-5 28-0
Fax +49 (0) 21 51-57 36 43
Email: info@alberdingk-boley.de
www.alberdingk-boley.de

Member of IVK

Company

Year of formation
1772

Size of workforce
450 (worldwide)

Ownership structure
family-owned

Subsidiaries
Alberdingk Boley Leuna GmbH, Leuna
Alberdingk Boley, Inc., Greensboro, USA,
Alberdingk Resins (Shenzhen) Co. Ltd.,
Shenzhen, China,
Alberdingk Italia S.r.l., Treviso Italy
Thai Castor Oil Industries Co., Ltd.,
Bangkok, Thailand

Sales channels
worldwide

Contact partners
Management:
Timm Wiegmann
Sales and Marketing

Application technology and sales:
Irene Tournee,
Technical Marketing Adhesives

Marketing Coatings:
Johannes Leibl, Manager Sales Dispersions

Range of Products

Raw materials
Polymers:
Polyurethane dispersions
Acrylate dispersions
Styrene acrylic dispersions
Vinylacetate copolymer dispersions
UV-curable dispersions

For applications in the field of
Adhesives:
Tapes, Labels,
Paper/Packaging
Construction
Automotive
Foils/flexible packaging

Coatings:
Wood/Furniture
Metal/Plastics
Construction including floors, walls and
ceilings
Textile/Leather
Film coatings
Primer
Barrier coatings for paper

ALFA Klebstoffe AG
Vor Eiche 10
CH-8197 Rafz
Phone +41 43 433 30 30
Fax +41 43 433 30 33
Email: info@alfa.swiss
www.alfa.swiss

Member of FKS

Company

Year of formation
1972

Size of workforce
63

Ownership structure
Family Private Limited (AG) Company

Subsidiaries
ALFA Adhesives, Inc. (Partner)
SIMALFA China Co. Ltd.

Sales channels
International distribution network

Contact partners
Management:
info@alfa.swiss

Application technology and sales:
info@alfa.swiss

Further information
ALFA Klebstoffe AG, an innovative family
business, focused on the development,
production and distribution of water-based
adhesives and hotmelts; therefore, ALFA
Klebstoffe AG offers a significant benefit
regarding the economic and ecologic design
of the gluing process of their customers.

Range of Products

Types of adhesives
Dispersion adhesives
Hot melt adhesives
Pressure-sensitive adhesives

For applications in the field of
Foam converting industry
Matresses production
Upholstery
Paper/packaging
Bookbinding/graphic design
Wood/furniture industry
Automotive industry, aviation industry
Hygiene

APM Technica AG
Max-Schmidheinystrasse 201
CH-9435 Heerbrugg
Phone +41 (0) 71 788 31 00
Fax +41 (0) 71 788 31 10
Email: info@apm.technica.com
www.apm-technica.com

Member of FKS

Company

Year of formation
2002

Size of workforce
120

Ownership structure
Private Shareholder

Subsidiaries
APM Technica Philippines, APM Technica GmbH, Abatech, Polyscience AG

Sales channels
Direct & Distributor

Contact partners
Management:
Andreas Hedinger, Giorgio Rossi

Application technology and sales:
APM Technica AG is the full-service provider in the field of adhesives and surface technology.

Further information
www.apm-technica.com

Range of Products

Types of adhesives
Reactive adhesives

Types of sealants
Polysulfide sealants
Silicone sealants
MS/SMP sealants

Equipment, Plant and Components
for surface pretreatment
for adhesive curing
adhesive curing and drying

For applications in the field of
Electronics
Mechanical engineering and equipment construction
Automotive industry, aviation industry
Medical and optic industry

Arakawa Europe GmbH
Düsseldorfer Straße 13
D-65760 Eschborn
Phone +49 (0) 6196-50383-0
Fax +49 (0) 6196-50383-10
Email: info@arakawaeurope.de
www.arakawaeurope.com

Member of IVK

Company

Year of formation
1998

Size of workforce
< 100

Managing partners
Arakawa Chemical Indstries Ltd.

Nominal capital
52.000,- €

Ownership structure
100 % affiliated company of
Arakawa Chemical Indstries Ltd.

Contact partners
Managing Director: Nobuyuki Fuke,
General Manager: Uwe Holland

Application technology and sales:
Dr. Ulrich Stoppmanns

Further information
Production and sale of hydrogenated
hydrocarbon resins, Rosin ester derivatives.

Range of Products

Raw materials
C9 Resins, Natural resins

For applications in the field of
Paper/Packaging
Bookbinding/Graphic design
Wood/Furniture industry
Automotive industry, Aviation industry
Adhesive tapes, Labels
Hygiene
Household, Recreation and Office
Cosmetics, Pharmaceutical industry

ARDEX Gmbh

Friedrich-Ebert-Straße 45
D-58453 Witten
Phone +49 (0) 2 30 26 64-0
Fax +49 (0) 2 30 26 64-3 75
Email: technik@ardex.de
www.ardex.de

Member of IVK, FCIO, VLK

Company

Year of formation
1949
Size of workforce
3,300
Ownership structure
Private
Subsidiaries worldwide
LUGATO GmbH & Co. KG, Germany
GUTJAHR Systemtechnik GmbH, Germany
ARDEX Schweiz AG, Switzerland
The W. W. Henry Company L.P., USA
ARDEX L.P., USA
ARDEX UK Ltd., Great Britain
Building Adhesives Limited, Great Britain
ARDEX Building Products Ireland Limited,
Ireland
ARDEX Australia Pty.. Ltd., Australia
ARDEX New Zealand Ltd., New Zealand
ARDEX Singapore Pte. Ltd., Singapore
QUICSEAL Construction Chemicals Ltd.,
Singapore
ARDEX Manufacturing SDN.BHD., Malaysia
ARDEX Taiwan Inc., Taiwan
ARDEX HONG KONG LIMITED, China
ARDEX (Shanghai) Co. Ltd., China
ARDEX Vietnam ARDEX Korea Inc., Korea
ARDEX Endura (INDIA) Pvt. Ltd., India
ARDEX Baustoff GmbH, Austria
ARDEX Epitöanyag Kereskedelmi Kft., Hungary
ARDEX Baustoff s.r.o., Czech Republic
ARDEX EOOD, Bulgaria
ARDEX Russia OOO, Russia
ARDEX Yapi Malzemeleri Ltd. Sti., Turkey
ARDEX s.r.l, Italy
ARDEX Romania s.r.l, Romania
DUNLOP Romania, Romania
ARDEX Middle East FZE, Dubai
ARDEX Skandinavia A/S, Denmark

Range of Products

Types of adhesives
Dispersion adhesives
Pressure-sensitive adhesives
Types of sealants
Acrylic sealants
Butyl sealants
PUR sealants
Silicone sealants
MS/SMP sealants
For applications in the field of
Construction industry, including floors,
walls and ceilings

ARDEX Skandinavia AS, Filial Norge, Norway
ARDEX-ARKI AB, Sweden
ARDEX OY, Finland
ARDEX Polska Sp.zo. o., Poland
ARDEX France S.A.S., France
ARDEX CEMENTO S.A., Spain
SEIRE Products S.L., Spain
ARDEX Cementos Mexicanos, Mexico
Wakol GmbH, Germany
Knopp Gruppe, Germany
DTA Australia Pty. Ltd., Australia
Nexus Australia Pty. Ltd., Australia
Ceramfix Argamassas E Rejuntes, Brazil
LOBA GmbH & Co. KG, Deutschland

Contact partners
Management:
Mark Eslamlooy (CEO)
Dr. Ulrich Dahlhoff
Dr. Hubert Motzet
Uwe Stockhausen
Dr. Markus Stolper

Application technology and sales:
Daniel Händle

ARLANXEO Deutschland GmbH

Chempark Dormagen, Building F41
Alte Heerstraße 2
D-41540 Dormagen
www.arlanxeo.com

Member of IVK

Company

ARLANXEO is one of the world's largest producers of synthetic rubber and a wholly owned subsidiary of Saudi Aramco, a leading producer of energy and chemicals. It develops, produces and markets high-performance rubbers with sales of around EUR 3 billion in 2019 and a presence at more than 12 production sites in 10 countries and 7 innovation centers around the world. Its products are used for a wide range of applications: from the automotive and tire industries to the electrical, construction and oil and gas industries. For more information, please visit www.arlanxeo.com. Follow us on LinkedIn.

Contact persons

Dr. Martin Schneider
Mobil: +49 175 31 23096
Email: martin.schneider@arlanxeo.com

Dr. Rainer Kalkofen
Mobil: +49 175 31 21249
Email: rainer.kalkofen@arlanxeo.com

Range of Products

Raw materials for the production of adhesives and sealants
Special grades from the product lines Baypren® (chloroprene rubber), Levamelt® (ethylene-vinyl acetate copolymers), X_Butyl® (butyl rubber) and Baymod® N, Krynac®, Perbunan® (all nitrile butadiene rubber) are tailor-made for the use in adhesive applications. The synthetic rubbers are offering unique properties regarding elasticity, polarity and tackiness, making them particular suitable for versatile applications in the adhesive and sealant industry:

- Baypren®: First choice for solvent-borne contact adhesive
- Levamelt®: Base polymer for pressure sensitive adhesives and modifier for structural adhesives and hot melts
- X_Butyl®: Basis for pressure sensitive adhesives and sealants
- Baymod® N, Krynac®, Perbunan®: Basis for contact adhesives and modifier to epoxy adhesives

artimelt AG

Wassermatte 1
CH-6210 Sursee
Phone +41 41 926 05 00
Email: info@artimelt.com
www.artimelt.com

Member of FKS

Company

Year of formation
2016

Ownership structure
artimelt is a family owned company in Switzerland.

Subsidiary, Representative
artimelt Inc., Tucker, GA 30084, USA
Formosa Nawonsuith Corp. (FNC), 11552 Taipei, Taiwan

Sales channels
Direct sales and agents

Managing director
Walter Stampfli

Contact partner
Christian Fischer
Phone: +41 41 926 05 35
Email: christian.fischer@artimelt.com

Further information
artimelt combines many years of experience with a high degree of innovation; for future-oriented and sustainable solutions. artimelt is always looking for something special and supports its customers on a partnership basis to them be competitive in their market.

Range of Products

Types of adhesives
Hot melt adhesives

For applications in the field of
Medical applications
Labels
Tapes
Packaging
Security systems
Building
Specialities

ASTORtec AG
Zürichstrasse 59
CH-8840 Einsiedeln
Phone +41 55 418 37 37
Fax +41 55 418 37 38
Email: info@astortec.ch
www.astortec.ch

Member of IVK

Company

Year of formation
1970

Size of workforce
45

Ownership structure
Private shareholders

Subsidiaries
ASTORPLAST Klebetechnik GmbH, POLYSCHAUM Packtechnik und Isoliermaterial GmbH

Sales channels
Direct sales and distributors

Contact partners
Management: Roland Leimbacher

Application technology and sales:
Guillaume Douard, Development Manager

Further information
ASTORtec AG was formed in 2019 from the merger of ASTORplast AG and Astorit AG, both companies with a heritage of over 50 years. The company is a Swiss SME with a focus on customer-specific solutions with the slogan: "bonds. seals. protects. - fits."

ASTORtec AG offers following solutions:
- MIXING of DEFINED FORMULATIONS for: batch size of 15 to 1'000 litres · reactive resins, adhesives, chemical raw materials · processing under ATEX (explosion protection) conditions · processing of UV-sensitive materials
- FILLING and CUSTOMIZED PACKAGING for: cartridges: 1-K, 2-K systems (1:1 to 1:10) · from small containers to IBC (1000 litres): tubes, cans, bottles, hobbock, drums, IBC · thin- to grease like products · wide range of standard packaging types · under ATEX (Ex-2) conditions possible · incl. labelling and proper handling of hazardous substances · worldwide delivery
- FORMULATING & DEVELOPING customer-specific modifications of standard-products: addition of fillers, colouring,

Range of Products

Types of adhesives
Hot melt adhesives; Reactive adhesives; Solvent-based adhesives; Dispersion adhesives; Pressure-sensitive adhesives

Types of sealants
Acrylic sealants; PUR sealants; Silicone sealants; MS/SMP sealants; Other

Raw materials
Additives; Fillers; Resins; Solvents; Polymers

Equipment, plant and components
For conveying, mixing, metering and for adhesive application; measuring and testing

For applications in the field of
Construction industry, including floors, walls and ceilings; Electronics; Mechanical engineering and equipment construction; Automotive industry, aviation industry; Adhesive tapes, labels

---

thixotropic treatment, diluting, accelerating, flexibilising
- SEALINGS robot applied direct sealings (RADS/FIPFG) · as contract manufacturing on injection moulded and sheet metal parts, incl. assembly steps · sealing tapes & cross coils of self-adhesive foams · die-cut parts made of foam and pressure sensitive tapes
- PRODUCT PROTECTION/PACKAGING distance pads made of cork, polymer foam, carton honey-comb · spacers made of foam · tapes/profiles for insulation and acoustic protection in construction
- QUALITY ASSURANCE & SERVICE laboratory to test the properties of adhesives · test certificates · application advise & aptitude tests
- TRADE of adhesives & chemical raw materials

ADHESIVE SYSTEMS

ATP adhesive systems AG
Sihleggstrasse 23
CH-8832 Wollerau
Phone +41 (0) 43 888 15 15
Fax +41 (0) 43 888 15 10
Email: info@atp-ag.com
www.atp-ag.com

Company

Year of formation
1989

Size of workforce
250

Managing partners
Managing Director:
Daniel Heini

Ownership structure
Privately owned

Sales channels
Direct sales channels
Graphic Distributors

Further information
ATP is a leading manufacturer of high quality technical tape solutions for the automotive, foam, graphic, label, semi-structural composite, building and construction industries. With our extensive technical and marketing knowledge, our passion for developing customer-focused solutions and our committed employees that go that little bit further, success with ATP is a given.
ATP is producing high quality single- and double-sided adhesive tapes on very modern coating machines in Germany since 1991. Using a broad range of adhesive and support materials, ATP produces single- and double-sided pressure sensitive adhesive tapes, transfer tapes and heat-seal films. The adhesive formulations are solvent free and are exclusively developed by ATP.

Range of Products

Types of adhesives
Dispersion adhesives
Pressure-sensitive adhesives
Self adhesive tapes
(Single- and double-sided tapes with different carriers)

For applications in the field of
Paper/packaging
Bookbinding/graphic design
Wood/furniture industry, including floors, walls and ceilings
Electronics
Automotive industry, aviation industry
Textile industry
Adhesive tapes, labels

ATP's production methods meet the most modern technological requirements. The products are developed and manufactured under a quality management system which is certified according to ISO 9001, ISO 14001, ISO 50001 and ISO/TS 16949.
ATP always strives to exceed customer expectations by developing customer aligned solutions which offer technical advantages, competitively and quickly.

Avebe U. A.

Prins Hendrikplein 20
NL-9641 GK Veendam
Phone +31 (0) 598 66 91 11
Fax +31 (0) 598 66 43 68
Email: info@avebe.com
www.avebe.com

Member of VLK

Company

Established in
1919

Employees
1,311

Ownership structure
Cooperation of farmers

Sales channels
Direct and through specialized
distributors worldwide

Further information
Avebe U. A. is an international Dutch starch
manufacturer located in the Netherlands and
produces starch products based on potato
starch and potato protein for use in food,
animal feed, paper, construction, textiles and
adhesives.

Range of Products

Raw materials
Dextrins and Starch based adhesives
Starch ethers

For applications in the field of
Paper and packaging
Paper sack adhesive
Tube winding adhesive
Remoistable envelope adhesive
Protective colloid in polyvinyl acetate
based dispersions
Water activated gummed tape adhesive
Wallpaper and bill posting adhesive
Additive – Rheology modifier for cement
and gypsum based mortars and tile
adhesives
Water purification

Avery Dennison Materials Group Europe

Sonnenwiesenstrasse 18
CH-8280 Kreuzlingen
Phone +41 (0) 71 68 68-100
Fax +41 (0) 71 68 68-181
www.averydennison.com

Member of IVK

Company

Year of formation
1984

Size of workforce
100 employees

Ownership structure
GmbH

Range of Products

Types of adhesives
Hot melt adhesives
Solvent-based adhesives
Dispersion adhesives
Pressure-sensitive adhesives

For applications in the field of
Adhesive tapes, labels

BASF SE
D-67056 Ludwigshafen
Phone +49 (0) 6 21-60-0
www.basf.com/adhesives
www.basf.com/pib

Member of IVK, VLK

Company

Year of formation
1865

Size of workforce
approximately 110,000 employees
(as of year end 2020)

Range of Products for Adhesives
1. Acrylic dispersions (water-based)
2. Acrylic hot melts (UV-curable)
3. Polyurethane dispersions (PUD)
4. Styrene-butadiene dispersions
5. Polyisobutene (PIB)
6. Polyvinylpyrrolidone (PVP)
7. Polyvinylether (PVE)
8. Auxiliaries:
   • Crosslinking agents
   • Defoamer
   • Thickener
   • Wetting agents
9. Additives
   • Antioxidants
   • Light stabilizer
   • Photoinitiators/Curing agents
   • Others

Email contacts
1 – 9: adhesives@basf.com
5: info-pib@basf.com

Range of Products

Comments
BASF is one of the world's leading manufacturers of raw materials and additives for adhesives.

With its technologies, BASF offers its customers an alternative to traditional fastening methods. The additive portfolio helps to improve products no matter which adhesive technology is used. BASF's polymer dispersions, UV curable hot melts and auxiliaries are an excellent solution for the production of high-quality self-adhesive products such as labels, tapes and films.

For highly sophisticated adhesives, BASF assists with know-how, reliability and safety based on decades of experience. BASF constantly develops and tests new products – according to customer requirements.

BASF also supplies a wide range of polyisobutenes (PIB) with low-, medium- and high molecular weight. Beyond other applications, Glissopal® (LM PIB) is used as tackifier to adjust stickiness of adhesive formulations. The applications for Oppanol® (MM and HM PIB) range from adhesives, sealants through to chewing gum base.

Die
Chemie**Versteher**

BCD Chemie GmbH
Schellerdamm 16
D-21079 Hamburg
Phone: +49 40 77173-0
Fax: +49 40 77173 2640
Email: info@bcd-chemie.de
www.bcd-chemie.de

Member of IVK

Company

Size of workforce
> 100 employees

Managing partners
Ronald Bolte (Chairman)
Mike Dudjan
Lothar Steuer

Range of Products

Raw materials
Additives: Dispersing agents, defoamer, adhesion promoter, hydrophobing agents, wetting agents, thickener, rheological additives, silanes, flame retardant agents

Resins: Thermoplastic acrylates, styrene acrylic and pure acrylic dispersions, vinyl acetate dispersions, polybutadien

Solvents: All kind of solvents like alcohols, esters, ketones, glycolethers, aliphatics, amines and many more

Polymers: Thermoplastic acrylates, styrene acrylic and pure acrylic dispersions, vinyl acetate dispersions, polybutadien

For applications in the field of
Paper/packaging
Bookbinding/graphic design
Wood/furniture industry
Construction industry, including floors, walls and ceilings
Electronics
Automotive industry, aviation industry
Textile industry
Adhesive tapes, labels
Hygiene
Household, recreation and office

BEARDOW ADAMS™

Unique Adhesives

Beardow Adams GmbH
Vilbeler Landstraße 20
D-60386 Frankfurt/M.
Phone +49 (0) 69-4 01 04-0
Fax +49 (0) 69-4 01 04-1 15
www.beardowadams.com

Member of IVK

Company

Year of formation
1875

Sales channels
direct and agencies

Contact partners
Sales and Marketing Management:
Janet Pohl

Marketing Assistant:
Suzie Daniel

Range of Products

Types of adhesives
Hot melt adhesives
Reactive adhesives
Dispersion adhesives
Casein, dextrin and starch adhesives
Pressure-sensitive adhesives

For applications in the field of
Paper converting/packaging/labelling
Bookbinding/graphic design
Wood/furniture industry
Electronics
Mechanical engineering and equipment
construction
Automotive industry
Adhesive tapes, labels

Bilgram Chemie GmbH

Torfweg 4
D-88356 Ostrach
Phone +49 (0) 7585 9312-0
Fax +49 (0) 7585 9312-94
Email: info@bilgram.de
www.bilgram.de

Member of IVK

Company

Year of formation
1971

Size of workforce
250

Ownership structure
family-owned

Sales channels
direct sales, wholesale trade, trade partners

Contact partners
Application technology:
Email: roland.opferkuch@bilgram.de
Sales:
Email: verena.leupolz@bilgram.de

Range of Products

Types of adhesives
PVAC – dispersions
Latexemulsions (natural rubber)
Urea/Formaldehyde resins (powder/liquid)
Hardener/Crosslinker for dispersions
Salthardener for U/F MUF resins

For applications in the field of
joint veneer adhesive
lamination of foils/paper/veneer
plywood (seatshells/handrails ...)
table tennis racket (glue/peeling ...)
reinforcement concrete, cement, mortar

Biesterfeld Spezialchemie GmbH
Ferdinandstraße 41
D-20095 Hamburg
Fax +49 (0) 40 32008-671
Email: m.liebenau@biesterfeld.com
www.biesterfeld.com

Company

Year of formation
1988

Size of workforce
262

Managing partners
Peter Wilkes, Thomas Arnold

Ownership structure
100% subsidiary of Biesterfeld AG

Subsidiaries
in more than 20 countries throughout
Europe

Contact partners
Management:
Dr. Martin Liebenau, Business Manager
CASE, m.liebenau@biesterfeld.com

Range of Products

Types of adhesives
Hot melt adhesives
Reactive adhesives
Solvent-based adhesives
Dispersion adhesives
Vegetable adhesives, dextrin and starch
adhesives
Glutine glue
Pressure-sensitive adhesives

Types of sealants
Acrylic sealants, Butyl sealants
Polysulfide sealants, PUR sealants
Silicone sealants, MS/SMP sealants, Other

Raw materials
Additives: Antioxidants, CMC, Deaerators,
Defoamers, Dispersants, Emulsifiers, Epoxy
hardeners, Flame retardants, Flow and
Leveling Agents, PUR-catalysts, Rheology
modifiers, UV-stabilizers, Wetting agents,
Xantan Gum
Resins: Acrylic Resins, Polybutens, Polyester-
polyols, Polyetherpolyols, Pre-Polymers
Polymers: MS Polymers
Starch: Yellow and white dextrin, Starch
esters, Starch ethers

For applications in the field of
Paper/packaging
Bookbinding/graphic design
Wood/furniture industry
Construction industry, including floors,
walls and ceilings
Electronics
Textile industry
Adhesive tapes, labels

BOLTON ADHESIVES

Fax + 31 (0) 88 3 235 800
Email: info@boltonadhesives.com
www.boltonadhesives.com
www.bison.net, www.griffon

Bison International B.V.
Dr. A.F. Philipsstraat 9
NL-4462 EW Goes
Phone +31 (0) 88 3 235 700

Headquarter: Bolton Adhesives
Adriaan Volker Huis – 14th floor
Oostmaaslaan 67, NL-3063 AN Rotterdam

Member of IVK

Company

Year of formation
1938

Size of workforce
Bolton Adhesives: > 750

Subsidiaries
Bison International, Zaventem (B)
Griffon France S.a.r.l, Compiègne (F)
Productos Imedio S.A., Madrid (E)

Sales channels
Technical Trade, Hardwarestore, Food
Distribution, Paper-, Office- & Stationery
Trade, Drugstores, Departmentstores

Contact partners
Management:
Rob Uytdewillegen, Danny Witjes

Further information
Technology:
Wiebe van der Kerk, Mariska Grob,
Charlotte Janse, John van Duivendijk

Sales:
Professional Business: Egbert Willemse
DIY: Frank Heus

Range of Products

Types of adhesives
Hot melt adhesives
Reactive adhesives
Solvent-based adhesives
Dispersion adhesives
Pressure-sensitive adhesives

Types of sealants
Acrylic sealants
PUR sealants
Silicone sealants
MS/SMP sealants

For applications in the field of
Paper/packaging
Wood/furniture industry
Construction industry, including floors, walls
and ceilings
Electronics
Automotive industry, aviation industry
Adhesive tapes, labels
Household, recreation and office

BODO MÖLLER CHEMIE
Engineer chemistry

Bodo Möller Chemie GmbH
Senefelderstraße 176
D-63069 Offenbach am Main
Phone +49 (0) 69-83 83 26-0
Fax +49 (0) 69-83 83 26-199
Email: info@bm-chemie.de
www.bm-chemie.com

Company

Year of formation
1974

Employees
Over 200

Managing partners
Korinna Möller-Boxberger, Frank Haug, Jürgen Rietschle

Ownership structure
Family owned

Regions
Germany, Austria, Slovenia, Switzerland, France, Benelux, United Kingdom, Ireland, Denmark, Sweden, Norway, Finland, Estonia, Poland, Lithuania, Latvia, Czech Republic, Slovakia, Hungary, Croatia, Romania, Spain, Russia, India, China, South Africa, Sub-Saharan Africa, Kenya, Egypt, Morocco, Middle East, Israel, USA and Mexico.

Sales channels
Own sales and logistics structures in Europe, Africa, Asia and America.

Contact partners
info@bm-chemie.de

Further information
Leading partner of global chemical companies, like Huntsman, Dow, DuPont, BASF and Henkel, with 45 years of experience in application engineering. Global distributor for high-performance epoxy resin, polyurethane and silicone-based adhesives, thermosetting plastics, pigments, additives, textile agents, dyes, electro casting resins, tooling and laminating resins and composites. In-house production facilities and laboratories for customer-specific formulations and application tests round off the range of services. Bodo Möller Chemie is certified for aviation and railway.

Range of Products

Types of adhesives
Epoxy resin adhesives, polyurethane adhesives, methacrylate adhesives, silicone adhesives, hot melt adhesives, reactive adhesives, solvent-based adhesives, dispersion adhesives, pressure-sensitive adhesives, MS polymers, polycondensation adhesives, UV-curing adhesives, spray adhesives, anaerobic adhesives, cyanoacrylates

Types of sealants
Acrylic sealants, butyl sealants, polysulfide sealants, PUR sealants, silicone sealants, MS/SMP sealants

Raw materials
Additives: stabilizers, antioxidants, rheology modifiers, tackifier, softener, thickener, dispersing agent, flame retardant, pigments, light stabilizers: HALS and UV stabilizers, crosslinkers

Fillers: barium sulphate, dolomite, kaolin, calcium carbonate, zinc, talc, aluminium oxide

Resins: acrylate dispersions, polyurethane dispersions, epoxy resins, rosin resins, reactive diluent, cobalt-free drying agent

Polymers: Formulated polymers EP, PU, PA

For applications in the field of
Paper/packaging, bookbinding/graphic design, wood/furniture industry, construction industry including floors, walls and ceilings, electronics, mechanical engineering and equipment construction, automotive industry, aviation industry, textile industry, sports and leisure, marine, adhesive tapes, labels, hygiene, household, recreation and office, adhesive applications for lightweight construction, composite bonding

Bona AB
Murmansgatan 130, Box 210 74
S-20021 Malmö
Phone +46 40 38 55 00, Fax +46 40 18 16 43
Email: bona@bona.com
Adhesive Production
Bona GmbH Deutschland
Jahnstraße 12, D-65549 Limburg
Phone +49 (0) 64 31-4 00 80

Member of IVK

Company

Range of Products

Year of formation
1919

Size of workforce
600

Ownership structure
privat owned stock cooperation

Subsidiaries
Austria: Bona Austria GmbH
Phone +43 662 66 19 43-0
Belgium: Bona NV
Phone +32 2 721 2759
Brazil: BonaKemi Pisos de Madeira
do Brasil Ltda
Phone +55 41 3233 5983
Bona AB Branch Panama
Phone +507 227 2799
Bona Trading (Shanghai) Co., Ltd
Phone +86 10 67 72 8301
Czech Republic & Slovak Republic:
Bona CR spol. s.r.o.
Phone CR +420 236 080 211
Phone SR +421 265 457 161
France: Bona France
Phone: +33 3 88 49 18 60
Germany
Bona Vertriebsges. mbH Deutschland
Phone +49 6431 4008 0
The Netherlands: Bona Benelux BV
Phone +31 23 542 1864
Poland: Bona-Polska Sp. z.o.o.
Phone +48 61 816 34 60/61
Romania: Bona S.r.l.
Phone +40 31 405 75 93

Types of adhesives
Reactive adhesives
Dispersion adhesives

For applications in the field of
Construction industry, for wood floors and
LVT's

Singapore: BonaFar East & Pacific Pte Ltd
Phone +65 6377 1158
Spain/Portugal: Bona Ibérica
Phone +34 916 825522
Italy: Biffignandi spa – Via Circonvallazione
Est 2/6 – Cassolnovo (PV) Italia
Phone 0381 920111
Hungary: M.L.S. Magyarország Kft. -
2310 Szigetszentmiklós, Sellő u. 8.
HUNGARY - Phone/Fax: (06 24) 525 400
United Kingdom: Bona Limited
Phone +44 1908 525 150
United States: Bona US
Phone +1 303 371 1411

Contact partners
Management
Dr. Thomas Brokamp (Production, R & D)

Product Management:
Torben Schuy, Thomas Hallberg

smart adhesives

Bostik GmbH
An der Bundesstraße 16
D-33829 Borgholzhausen
Phone +49 (0) 54 25-8 01-0
Email: info.germany@bostik.com
www.bostik.de

Member of IVK, FEICA

Company

Year of formation
1889

Size of workforce
400

Subsidiaries
MEM Bauchemie GmbH, Bostik Austria,
Bostik Technology GmbH,
Bostik Aerosols GmbH

Sales channels
Construction distribution, DIY,
industry

Contact partners
Management:
Olaf Memmen, Managing Director
Richard Riepe, Business Manager Construction
Norbert Uniatowski, Business Manager FlexLam
Frank Mende, R&D Director
Stephan Rensch, Supply Chain Director
Dr. Michael Nitsche, Production Director

Further information
With annual sales of € 2.1 billion, the company employs 6,000 people and has a presence
in more than 50 countries. For the latest information, visit www.bostik.com.

Range of Products

Types of adhesives
Hot melt adhesives
Reactive adhesives
Solvent-based adhesives
Dispersion adhesives
Vegetable adhesives, dextrin and
starch adhesives
Pressure-sensitive adhesives
Polymer modified binders
SMP adhesives

Types of sealants
Acrylic sealants
Butyl sealants
PUR sealants
Silicone sealants
MS/SMP sealants

Raw materials
Additives, Fillers, Resins, Solvents
Polymers, Starch

For applications in the field of
Paper/packaging
Bookbinding/graphic design
Wood/furniture industry
Construction industry, including floors,
walls and ceilings
Mechanical engineering and equipment
construction
Automotive industry, aviation industry
Textile industry
Hygiene
Household, recreation and office
Flexible Laminating

Botament
Systembaustoffe
Gmbh & Co. KG

IZ NÖ-Süd Straße 7, Objekt 58C
A-2355 Wiener Neudorf
Phone +43 (0) 22 36-38 70 25
Email: info@botament.at
www.botament.com

Member of IVK, FCIO

Company

Year of formation
1993

Size of workforce
15

Ownership structure
GmbH & Co. KG

Sales channels
wholesale

Contact partners
Sales manager:
Prok. Ing. Peter Kiermayr

Application technology and sales:
Karl Prickl

CEO:
DI (FH) Markus Weinzierl

Range of Products

Types of adhesives
Reactive adhesives
Dispersion adhesives
Tile adhesives/natural stone adhesives

Types of sealants
Silicone sealants

For applications in the field of
Construction industry, including floors walls
and ceilings

Brenntag GmbH
Messeallee 11
45131 Essen
Phone +49 (0) 201 / 64 96 - 0
Email: construction@brenntag.de
www.brenntag-gmbh.de

Member of IVK

Company

Year of formation
1874

Size of workforce
1,200

Managing partners
Dr. Colin von Ettingshausen
Thomas Langer
Marc Porwol

Nominal capital
154.5 Mio. Euro (Brenntag AG)

Ownership structure
Listed on the stock exchange (Brenntag AG)

Contact partners
Management:
Alain Kavafyan

Application technology and sales:
Michael Hesselmann
Markus Wolff

Range of Products

Raw materials
Additives:
Accelerators, Adhesion Promoters, Antioxidants, Biocides, Catalysts, Defoamers, Dispersing Agents, Matting Agents, Plasticizers, Polyether Amines, Rheology Modifiers, Silanes, Surfactants, Thickeners, UV-Stabilisers, Molecular Sieves

Resins:
Acrylic Dispersions and Resins, Acrylic Monomers, Styrene Acrylics
Epoxy Resins incl. Curing Agents, Reactive Diluents and Modifiers,
Hydrocarbon Resins
PU-Systems: Polyols and Isocyanates (aromatic and aliphatic),
Prepolymers, PUR-Dispersions
Silicone Resins and Emulsions

Solvents: all kinds of solvents
Specialty Amines and Catalysts

Polymers: PMMA, Silicones

For applications in the field of
Paper/packaging
Bookbinding/graphic design
Wood/furniture industry
Construction industry, including floors, walls and ceilings
Electronics
Automotive industry, aviation industry
Textile industry
Adhesive tapes, labels
Composites

ΙΙ BÜHNEN

Bühnen GmbH & Co. KG
Hinterm Sielhof 25
D-28277 Bremen
Phone +49 (0) 4 21-51 20-0
Fax +49 (0) 4 21-51 20-2 60
Email: info@buehnen.de
www.buehnen.de

Member of IVK

Company

Year of formation
1922

Size of workforce
102

Ownership structure
Private ownership

Subsidiaries
BÜHNEN Polska Sp. z o. o.
BÜHNEN B. V., NL
BÜHNEN, AT

Contact person
Managing Director & Shareholder:
Bert Gausepohl

Managing Director & Shareholder:
Jan-Hendrik Hunke

Marketing:
Heike Lau

Sales GER, AT, CH, NL, PL, Intl.
Jan-Hendrik Hunke

Range of Products

Distribution channels
Direct Sales, Distributors

Hot Melt Adhesives
The product range includes a
variety of different hot melt adhesives
for almost every application.
Available bases:
EVA, PO, POR, PA, PSA, PUR, Acrylate.
Available shapes:
slugs, sticks, granules, pillows, blocks,
cartridges, barrels, drums, bags.

Application Technology
Hot melt tank applicator systems with
piston pump or gear pump, PUR- and POR-
hot melt tank systems, PUR- and POR-bulk
unloader, hand guns for spray and bead
application, application heads for bead,
slot, spray and dot application and special
application heads for individual customer
requirements, hand-operated glue applicators, PUR- and POR glue applicators, wide
range of application accessories, customer-
oriented application, solutions.

Applications
Automotive, Packaging, Display Manufac-
turing, Electronic Industry, Filter Industry,
Shoe Industry, Foamplastic and Textile
Industry, Case Industry, Construction
Industry, Florists, Wood-Processing and
Furniture Industry, Labelling Industry.

BYK
Abelstraße 45
D-46483 Wesel
Phone +49 (0) 281-670-0
Fax +49 (0) 281-6 57 35
Email: info@byk.com
www.byk.com

Member of IVK

Company

Year of formation
1962

Size of workforce
More than 2,300 people worldwide

Managing partners
Dr. Tammo Boinowitz (CEO),
Alison Avery (CFO & NAFTA),
Gerd Judith (Asia),
Matthias Kramer (Production)

Ownership structure
BYK is a member of ALTANA, Germany

Subsidiaries
BYK-Chemie (Germany), BYK (Brazil), BYK Additives
(China), BYK (India), BYK Japan (Japan), BYK Korea
(Korea), BYK Netherlands (Netherlands),
BYK Chemie de México (México), BYK Asia Pacific
(Singapore, Taiwan, Thailand and Vietnam),
BYK Additives (United Kingdom), BYK (U.A.E.),
BYK USA (USA)
- Warehouses and representations in
  > 100 countries and regions
- Technical Service Labs in Germany, Brazil, China,
  India, Japan, Korea, México, The Netherlands,
  Singapore, U.A.E., United Kingdom and in the USA
- Production Sites in Germany, China, The
  Netherlands, United Kingdom and in the USA

Sales channels
Worldwide – direct (BYK) and indirect
(agents and distributors)

Close to the customer
BYK places great importance on being close to its
customers and remaining in constant dialog with
them. This is one of the reasons why the company is
represented in more than 100 countries and regions
around the globe. In over 30 technical service
laboratories, BYK offers customers and application
engineers support for concrete questions.

Range of Products

Raw materials
Additives: Wetting and dispersing additives, rheological additives (PU thickener, organically modified
clays, synthetic and natural layered silicates),
defoamers and air release agents, additives to
improve substrate wetting, surface slip and levelling, UV-absorbers, wax additives, anti-blocking
additives, anti-static additives, water scavengers,
adhesion promoters

For applications in the field of
Paper/packaging
Bookbinding/graphic design
Wood/furniture industry
Construction industry including floors,
walls and ceilings
Electronics
Sealants
Automotive industry, aviation industry
Textile industry
Adhesive tapes, labels
Hygiene
Household, recreation and office

Contact partners for your customers
Mr. Tobias Austermann,
Email: Tobias.Austermann@altana.com
Phone: +49 281-670-28128

Further information about your company
BYK is a leading global supplier of specialty chemicals. The company's innovative additives and dif-
ferentiated solutions optimize product and material
properties as well as production and application
processes. Amongst other things, BYK additives
improve scratch resistance and surface gloss, the
mechanical strength or flow behavior of materials,
and properties such as UV- and light stability or
flame retardancy.

Byla GmbH

Industriestraße 12
D-65594 Runkel
Phone +49 (0) 64 82-91 20-0
Fax +49 (0) 64 82-91 20-11
Email: contact@byla.de
www.byla.de

Member of IVK

Company

Year of formation
1975

Nominal capital
90,000 €

Sales channels
Worldwide

Range of Products

Types of adhesives
Reactive adhesives

For applications in the field of
Wood/furniture industry
Electronics
Mechanical engineering and equipment construction
Automotive industry, aviation industry

certoplast
Technische Klebebänder
GmbH

Müngstener Straße 10
D-42285 Wuppertal
Phone +49 (0) 20 2-2 55 48-0
Fax +49 (0) 20 2-2 55 48-48
Email: verkauf@certoplast.com
www.certoplast.com

Member of IVK
Member of Afera

Company

Year of formation
1991

Size of workforce
90

Managing partners
Dr. R. Rambusch
Dr. A. Hohmann

Subsidiaries
certoplast (Suzhou) Co., Ltd., China
certoplast North America Inc., Las Cruses

Contact partners
Management:
Dr. R. Rambusch (General manager)
Dr. A. Hohmann (General manager)

Application technology and sales:
Dr. Andreas Hohmann (Sales manager)

Range of Products

Adhesive tapes based on
Hot melt adhesives systems
Dispersion adhesives
UV-acrylic adhesives

Equipment, plant and components
for adhesive curing
adhesive curing and drying

For applications in the field of
Construction industry, including floors,
walls and ceilings
Electronics
Automotive industry, aviation industry
Adhesive tapes, labels

We create chemistry

expect more ⊕

Chemetall GmbH
Trakehner Straße 3
D-60487 Frankfurt/M.
Phone +49 (0) 69-71 65-0
Fax +49 (0) 69-71 65-29 36
www.chemetall.com

Member of IVK

Company

Year of formation
1982

Managing partners
Board of Management:
Christophe Cazabeau

Ownership structure
A Global Business Unit of BASF's
Coatings division

Subsidiaries
> 40 worldwide

Sales channels
CM subsidiaries and specific distributors

Contact partners
Management:
Thomas Willems

Application technology and sales:
Ralph Hecktor
Phone +49 (0) 69-71 65-24 46

Further information
See website: www.chemetall.com
Certification to ISO 9001, EN 9100,
ISO 14001

Range of Products

Types of adhesives
Hot melt adhesives
Reactive adhesives

Types of sealants
Polysulfide sealants
PUR sealants
Other epoxy

For applications in the field of
Electronics
Mechanical engineering and equipment
construction
Automotive industry, Aviation industry

Chemische Fabrik Budenheim KG

Rheinstraße 27
D-55257 Budenheim
Phone +49 (0) 6139-89-0
Email: info@budenheim.com
www.budenheim.com

Member of IVK

Company

Year of formation
1908

Size of workforce
1.000

Managing partners
Dr. Harald Schaub
Dr. Stefan Lihl

Ownership structure
Belongs to the Oetker Group, privately owned

Contact partners
Management:
Email: coatings@budenheim.com

Applications technology and sales:
Email: coatings@budenheim.com

Furher information
www.budenheim.com/clip4coatings

Range of Products

Raw materials
Additives
Fillers
Flame Retardants

For applications in the field of
Wood/furniture industry
Construction industry, including floors, walls and ceilings
Automotive industry, aviation industry
Textile industry

**SMART CHEMISTRY
WITH CHARACTER.**

CHT Germany GmbH
Bismarckstraße 102
D-72072 Tübingen
Phone +49 7071 154-0
Fax +49 7071 154-290
Email: info@cht.com
www.cht.com

Member of IVK

Company

Year of formation
1953

Size of workforce
2,200 worldwide

Sales channels
More than 26 CHT affiliates and agencies
worldwide

Management
Dr. Frank Naumann (CEO)
Dr. Bernhard Hettich (COO)
Axel Breitling (CFO)

Contact partners
Application technology:
Dennis Seitzer (R&D)

Range of Products

Types of adhesives
Acrylic adhesives
Dispersion adhesives
High solid adhesives
Powder adhesives
Pressure-sensitive adhesives
PUR adhesives
Reactive adhesives
Silicone adhesives
Solvent-based adhesives
Thermo-activated adhesives

Types of sealants
Silicone sealants

Raw materials
Acrylic dispersions
LSR silicones
PU dispersions
RTV-1 / RTV-2

Additives:
Adhesion promoters and primers
Crosslinkers, chain extenders
Release agents
Rheology additives and thickeners

For applications in the field of
Automotive industry, aviation industry
Construction industry
Electronics
Flock
Mechanical engineering and equipment
Mould making
Paper and packaging
Pressure sensitive tapes and labels
Textile industry/technical textiles

CnP Polymer GmbH

Schulteßdamm 58
D-22391 Hamburg
Phone +49 (0) 40-53 69 55 01
Fax +49 (0) 40-53 69 55 03
Email: info@cnppolymer.de
www.cnppolymer.de

Member of IVK

Company

Year of formation
1999

Ownership structure
private

Sales channels
own sales force

Contact
Christoph Niemeyer

Range of Products

Raw materials
Polymers:
SIS, SBS, SEBS, SSBR
PIB

Resins:
Hydrocarbon resins

For applications in the field of
Paper/packaging
Bookbinding/graphic design
Wood/furniture industry
Construction industry, including floors, walls and ceilings
Automotive industry, aviation industry
Textile industry
Adhesive tapes, Labels
Hygiene
Household, recreation and office

Coim Deutschland GmbH
Novacote Flexpack Division
Schnackenburgallee 62
D-22525 Hamburg
Phone +49 (0) 40-85 31 03-0
Fax +49 (0) 40-85 31 03-69
Email: info@coimgroup.com
www.coimgroup.com

Member of IVK

Company

Year of formation
as COIM Group in 1962

Ownership structure
private owned company

Subsidiaries
COIM operates through a network of production sites, commercial companies and agencies located all over the world

Sales channels
The Novacote Division is part of the COIM Group, dedicated to developing and supplying adhesives and coatings, mainly for the Flexible Packaging market.

Contact partners
Management:
Frank Rheinisch

Application technology:
Oswald Watterott

Sales:
Joerg Kiewitt

Further information
The Novacote Division is part of the Coim Group, dedicated to developing and supplying adhesives and coatings, mainly for the Flexible Packaging market. During recent years the Novacote Division grew rapidly both in terms of Business and Organization. With respect to the global organization the Novacote Technology Center is located in Hamburg, Germany as R&D Centre for Packaging.

Range of Products

Types of adhesives
Reactive adhesives
Solvent-based adhesives
Dispersion adhesives

Types of sealants
Acrylic sealants
Other

Raw materials
Resins
Polymers

For applications in the field of
Paper/packaging
Adhesive tapes, labels
Hygiene
Insulation materials
Solar panels

Collall B.V.

Electronicaweg 6
NL – 9503 EX Stadskanaal
Phone +31 (0) 599-65 21 90
Fax +31 (0) 599-65 21 91
Email: info@collall.nl
www.collall.nl

Member of VLK

Company

Year of formation
1949

Size of workforce
25

Ownership structure
family-owned

Contact partners
Management:
Patrick van Rhijn

Range of Products

Types of adhesives
Solvent-based adhesives
Dispersion Adhesives
Vegetable adhesives, dextrin and
starch adhesives

For applications in the field of
Household, hobby and office
Bookbinding/graphis design
Wood/furniture industry

Additional
supplier of various creative materials
for school and hobby

Collano AG
CH-6203 Sempach Station
Phone +41 41 469 92 75
Email: info@collano.com
www.collano.com

Member of FKS

Company

Year of formation
1947

Size of workforce
20

Sales channels
Direct sales or distributors

Contact partners
Management:
Mike Gabriel

Sales:
Gianni Horber
Phone: +41 41 469 92 75

Range of Products

Types of adhesives
Hot melt adhesives
Reactive adhesives
Dispersion adhesives

For applications in the field of
Wood/furniture industry
Construction industry
Transportation industry

Coroplast Fritz Müller GmbH & Co. KG
Wittener Straße 271
D-42279 Wuppertal
Phone +49 (0) 2 02-26 81-0
Fax +49 (0) 2 02-26 81-3 80
www.coroplast-tape.com

Member of IVK

Company

Year of formation
1928

Size of workforce (Coroplast Group)
7,000

Sales channels
Wholesale and industry

Contact partner
M. Söhngen

Range of Products

For applications in the field of
Paper/packaging
Wood/furniture industry
Construction industry, including floors, walls and ceilings
Electronics
Mechanical engineering and equipment construction
Automotive industry, aviation industry
Household, recreation and office

Covestro AG

D-51373 Leverkusen
Phone +49 (0) 214-6009 3080
Email: adhesives@covestro.com
www.adhesives.covestro.com

Member of IVK

Company

Year of formation
2015

Size of workforce
16,500 (31. December 2020)

Contact partners
Marketing Europe/Business Development
Phone +49 (0) 214-6009 3080
Email: adhesives@covestro.com
www.adhesives.covestro.com

Range of Products

Raw materials
Water-based acrylic resins (Bayhydrol® A, NeoCryl®, Decovery®)
Water-based urethane resins (Bayhydrol® UV, NeoPac™, NeoRez®)
Water-based urethane/acrylic hybrid resins (NeoPac™)
Polyurethane-Dispersions (Dispercoll® U)
Hydroxylpolyurethanes (Desmocoll®, Desmomelt®)
Polyisocyanates (Desmodur®)
Isocyanate-Prepolymers (Desmocap®, Desmodur®, Desmoseal®)
Silanterminated Polyurethanes (Desmoseal® S)
Polyesterpolyols (Baycoll®)
Polyetherpolyols (Desmophen®, Acclaim®)
Polychloroprene-Dispersions (Dispercoll® C)
Halogenated Polyisoprenes (Pergut®)
Silicon dioxide-nanoparticle dispersions (Dispercoll® S)

CTA GmbH
Voithstraße 1
D-71640 Ludwigsburg
Phone +49 (0) 71 41 - 29 99 16 - 0
Email: info@cta-gmbh.de
www.cta-gmbh.de

Member of IVK

CHEMIE | TECHNIK | ABFÜLLUNG | SEIT 1899

Company

Year of formation
2005

Size of workforce
200

Ownership structure
Member of Tubex Holding GmbH

Sales channels
Direct

Contact partners
Management:
Hans-Dieter Worch

Sales:
Franco Menchetti

Further information
The core business of CTA GmbH is a wide range of contract services in the fields of:
- **Product manufacturing** which includes the developing or improvement of composites and the mixing of products according
  to customers recipes and parameters
- **The filling** of almost all viscous fluid chemical products into various kinds of primary and secondary packaging – like all kind of tubes, cartridges, bottles, pouches, cartons, blister packs and others – which are released and approved to the best suitability of the product and its adequate packaging.

Range of Products

Types of adhesives
Solvent-based adhesives
Dispersion adhesives
1K isocyanate-based adhesives
1K and 2K epoxy-based adhesives
Liability adhesives
Fillers
MS polymer-based adhesives
Silane-based adhesives
Silicone

For applications in the field of
Paper/packaging
Wood/furniture industry
Construction industry, including floors, walls and ceilings
Electronics
Wind energy
Mechanical engineering and equipment construction
Automotive industry, aviation industry
Household, recreation and office

- **Developing** the appropriate packaging in accordance to the needs of the product and its marketing aspects.
- **The Packaging** assembling to the point of sale
- **Supply and distribution logistics** offers a full contract service package to different branches of industries and businesses.

H.B. Fuller | Cyberbond CB

Cyberbond Europe GmbH – A H.B. Fuller Company
Werner-von-Siemens-Straße 2, D-31515 Wunstorf
Phone +49 (0) 50 31-95 66-0, Fax +49 (0) 50 31-95 66-26
Email: info@cyberbond.de, www.cyberbond.eu

Member of IVK

Company

Year of formation
1999

Size of workforce
24

Managing partners
Ulrich Lipper, Holger Bleich, James East,
Robert Martsching

Nominal capital
50,000 EUR

Ownership structure
H.B. Fuller

Subsidiaries
Cyberbond France SARL, France
Cyberbond Iberia, Spain
Cyberbond CS s. r. o., Czech Republic

Sales channels
Direct to the industry and via exclusive
nationwide distributors as well as special
Private Label accounts

Managing partners
Holger Bleich, Gert Heckmann,
James East, Robert Martsching

Contact partners
Marketing and sales: Holger Bleich
Application technology: Nora Grotstück

Further information
Cyberbond – The Power of Adhesive
Information
IATF 16949
ISO 13485
ISO 9001
ISO 14001

Range of Products

Offered adhesives
Cyanoacrylates
Anaerobic Adhesives and Sealants
UV and Light Curing Adhesives
MMA 2-K adhesives
Additional programme consisting of:
Primers, Activators, D-Bonders and
Dosing Aids

Dosing equipment
LINOP Modular Dosing System for
1K Reactive Adhesives
LINOP UV LED Curing System

Products are used in
Automotive and automotive sub supplier
industry
Electronic industry
Aviation industry
Elastomer/plastic/metal working industry
Machine tool industry
Medical industry
Shoe industry
DIY, Hobby and office
Wood/furniture industry

DEKA
Kleben & Dichten
GmbH (Dekalin®)

Gartenstraße 4
D-63691 Ranstadt
Phone +49 (0) 60 41-82 03 80
Fax +49 (0) 60 41-82 12 22
Email: info@dekalin.de
www.dekalin.de

Member of IVK

Company

Year of formation
1907 DEKALIN
1999 DEKA

Size of workforce
together > 125

Ownership structure
family owned

Sales channels
industry, manufacturing,
technical wholesalers

Contact partners
Management:
Michael Windecker

Range of Products

Types of adhesives
Solvent-based adhesives
Dispersion adhesives

Types of sealants
Butyl sealants
PUR sealants
MS/SMP sealants

For applications in the field of
Wood/furniture industry
Construction industry, including floors,
walls and ceilings
Mechanical engineering and equipment
construction
Automotive industry, aviation industry
Household, recreation and office
Caravan, camper and mobile home,
HVAC, Air duct sealing

Delo Industrial Adhesives

DELO-Allee 1
D-86949 Windach
Phone +49 (0) 81 93-99 00-0
Fax +49 (0) 81 93-99 00-1 44
Email: info@DELO.de
www.DELO-adhesives.com

Member of IVK

Company

Year of formation
1961

Size of workforce
820

Managing board
Dr. Wolf-Dietrich Herold
Sabine Herold
Robert Saller

Subsidiaries
Subsidiaries in China, Japan, Malaysia, Singapore and the USA, representative offices and distributors in numerous other countries

Sales channels
Traders and direct

Contact partners
Application technology and sales:
Robert Saller, Managing director

Further information
DELO is a leading manufacturer of high-tech adhesives and other multifunctional materials as well as the corresponding dispensing and curing equipment. The company's products are mainly used in the automotive, consumer, and industrial electronics industries. Customers include Bosch, Daimler, Huawei, Osram, Siemens, and Sony. DELO achieved a turnover of 167 million euros in the financial year 2021 and is headquartered in Windach near Munich, Germany.

Range of Products

Types of adhesives
Dual-curing adhesives
Light-curing and light-activated acrylates and epoxies
One- and two-component epoxy resins
Electrically conductive adhesives
Methacrylates
Polyurethanes
Cyanoacrylates
Functional materials for optics, batteries and 3D printing

Types of sealants
Acrylic sealants
PUR sealants
Silicone sealants
Other

Equipment, plant and components
LED UV curing lamps (spot and area)
Jet-valves for micro-dispensing
Flexible foil cartridges for bubble-free dispensing

For applications in the field of
Automotive
Consumer and industrial electronics
Optics and optoelectronics
Mechanical engineering
Semiconductor
Aviation

Distona AG

Hauptplatz 5
CH-8640 Rapperswil
Phone +41 (0) 55 533 00 50
Fax +41 (0) 55 533 00 51
Email: info@distona.ch
www.www.distona.ch

Member of FKS

Company

Year of formation
2014

Size of workforce
5 employees

Managing partners
Daniel Altorfer
David Nipkow

Range of Products

Raw materials
Additives:
Stabilizers, biocides, flame retardants,
silane, siloxane, rheology modifiers

Fillers:
Porous Glas Powder, Bentonite, Kaolinite,
Chalk, Talcum, Zeolite

Resins:
Epoxide Technologies, Isocyanates, Polyols,
Acrylic monomers

Solvents Bio-solvents, conventional solvents

DKSH GmbH
Baumwall 3
20459 Hamburg
Phone +49 (0) 40-37 47-340
Fax +49 (0) 40-492 190 28
Email: info.ham@dksh.com
www.dksh.de

Member of IVK

Company

Year of formation
1972

Size of workforce
48

Managing Director
Thomas Sul

Ownership structure
DKSH Group

Subsidiaries of DKSH Group
32,450 employees/net sales of CHF 10.7
billion in 2020

Sales channels
Own sales force

Sales Manager
Sven Thomas

Further information
At DKSH, our purpose is to enrich people's lives. For more than 150 years, we have been delivering growth for companies in Asia and beyond across our Business Units Healthcare, Consumer Goods, Performance Materials and Technology. As a leading Market Expansion Services provider, we offer sourcing, market insights, marketing and sales, eCommerce, distribution and logistics as well as after-sales services. Listed on the SIX Swiss Exchange, DKSH operates in 36 markets with 32,450 specialists, generating net sales of CHF 10.7 billion in 2020. The DKSH Business Unit Performance Materials distributes specialty chemicals and ingredients for food, pharmaceutical, perso-

Range of Products

Raw materials
Wide range of specialties for Epoxi and PU
Aliphatic isocyanates
High-molecular weight co-polyesters
Liquid isoprene rubbers
Moisture scavengers (PTSI)
Oxetanic reactive diluents and plasticizers
Resins: Co- Polyester, Vinyl, Polyamide-Imide, Acrylic, Ketone, Oxetanic Resins
Heat seal lacquers
Adhesion promoters and primers Polyolefin, Polyolefindispersions, Polyester Lacquers
Plasticizers
Electrical and thermal conductive additives

For applications in the field of
Paper/Pharma/Food packaging Bookbinding/
graphic design Wood/furniture industry
Construction industry, including floors, walls and ceilings
Electronics
Automotive industry, aviation industry
Adhesive tapes, labels
Hygiene

nal care and various industrial applications. With 48 innovation centers and regulatory support worldwide, we create cutting-edge formulations that comply with local regulations.

Drei Bond GmbH
Carl-Zeiss-Ring 13
85737 Ismaning, Germany
Phone +49 (0) 89-962427 0
Fax +49 (0) 89-962427 19
Email: info@dreibond.de
www.dreibond.de

Member of IVK

Company

Year of formation
1979

Number of employees
47

Partners
Drei Bond Holding GmbH

Share capital
€ 50,618

Subsidiaries
Drei Bond Polska sp. z o.o. in Kraków

Distribution channels
Directly to the automotive industry
(OEM + tier 1/tier 2); indirectly
via trading partners as well as select private
label business

Contacts
Management:
Mr. Thomas Brandl

Application engineering, adhesive and
sealants:
Johanna Storm, Dr.Florian Menk

Application engineering, metering technology:
Sebastian Schmidt, Marko Hein

Adhesive and sealant sales:
Thomas Hellstern, Adrian Frey, Stephan Knorz

Metering technology sales:
Sebastian Schmidt, Marko Hein

Additional information
Drei Bond is certified according to ISO 9001-
2015 and ISO 14001-2015

Range of Products

Types of adhesives/sealants
- Cyanoacrylate adhesives
- Anaerobic adhesives and sealants
- UV-light curing adhesives
- 1C/2C epoxy adhesives
- 2C MMA adhesives
- 1 C + 2 C MS hybrid adhesives and sealants
- 1C synthetic adhesives and sealants
- 1C silicone sealants

Complementary products:
- Activators, primers, cleaners

Equipment, systems and components
- Drei Bond Compact metering systems →
  semi-automatic application of adhesives and
  sealants, greases and oils
  Metering technology: pressure/time and
  volumetric
- Drei Bond Inline metering systems →fully
  automated application of adhesives and
  sealants, greases and oils
  Metering technology: pressure/time and
  volumetric
- Drei Bond metering components:
  Container systems: tanks, cartridges, drum
  pumps
  Metering valves: progressive cavity pumps,
  diaphragm valves, pinch valves, spray
  valves, rotor spray

For applications in the following fields
- Automotive industry/automotive suppliers
- Electronics industry
- Elastomer/plastics/metal processing
- Mechanical and apparatus engineering
- Engine and gear manufacturing
- Enclosure manufacturing (metal and plastic)

DUNLOP TECH GmbH

Offenbacher Landstraße 8
D-63456 Hanau
Phone: +49 (0) 6181 9394-0
Fax: +49 (0) 6181 9394-553
Email: info@dunloptech.de
www.dunloptech.com

Member of IVK

Company

Year of formation
1997

Size of workforce
Internal 63 employees
Externally 95 employees

Managing partners
Sumitomo Rubber Industries Ltd. Kobe,
Japan (100%)

Nominal capital
19 Millionen €

Turnover
€ 50 million (2020)

Management
Bernd Schuchhardt
Hirokazu Nishimori

Contacts
Lars Biesenbach,
Head of Sales Innovative Products

Dr. Angel Jimenez,
Head of Adhesives & IMS Development

Distribution channel
Direct and indirect sales
(Industry and Distribution)

Range of Products

Types of adhesives
Latex-based adhesive made from sustainable
raw materials which is 99 % biodegradable

For applications in the field of
Interior for floors and walls

Further products
Deflation warning system (DWS), indirect,
software-based for passenger cars - automotive industry (OE).

Tire mobility Kits (IMS) for passenger cars
and light trucks - automotive industry (OE,
aftermarket), caravaning, bicycle and two-
wheeler industry.

DuPont Deutschland

Hugenottenallee 175
D-63263 Neu-Isenburg
Phone +49 (0)6102-18-0
Fax +49 (0)6102-18-1224
Email: ti.comms@dupont.com
www.DuPont.com/mobility

Member of IVK

Company

Founding year
1802 (DuPont)

Size of the workforce
5,600 employees, Mobility & Materials

Ownership
Mobility & Materials is a division of DuPont de Nemours, Inc.

Contact Person
Dr. Andreas Lutz, Global Technology Leader
Thorsten Schmidt, EMEA Regional Commercial Leader

Additional Information
DuPont Mobility & Materials (M&M) delivers a broad range of technology-based products and solutions to the automotive, electronics, industrial, consumer, medical, photovoltaic and telecom markets. DuPont M&M partners with customers to drive innovation by utilizing its expertise and knowledge in polymer and materials science. DuPont M&M works with customers throughout the value chain to enable material systems solutions for demanding applications and environments. For additional information about DuPont Mobility & Materials, visit www.dupont.com/mobility

Range of Products

Adhesive types & applications
- BETAFORCE™ TC and BETATECH™ thermal interface materials that help support battery thermal management during hybrid/electric vehicle charging and operation
- BETAFORCE™ multi-material bonding adhesives for vehicle body structure bonding, and battery sealing and assembly
- BETAMATE™ structural adhesives for vehicle body structure and battery bonding to support crash durability and lighter weight vehicle structures
- BETASEAL™ glass bonding adhesives that enhance vehicle structure for OEM installation and aftermarket repair of windscreens, backlites, panoramic, and stationary glass
- MEGUM™ and THIXON™ for rubber-to-substrate bonding

Dymax Europe GmbH
Kasteler Straße 45
D-65203 Wiesbaden
Phone +49 (0) 611-962 7900
Fax +49 (0) 611-962 9440
Email: info_DE@dymax.com
https://de.dymax.com

Member of IVK

Company

Year of formation
1995

Size of workforce
450 • worldwide

Ownership structure
Dymax Corporation, USA

Sales channels
Direct and Distributors

Contact partners
Managing Director:
Martin Senger

Technical Manager:
Dr. Thérèse Hémery, Wolfgang Lorscheider

Further information
At Dymax we combine our product offering of oligomers, adhesives, coatings, dispensing systems, and curing equipment with our expert knowledge of light-cure technology.

Range of Products

Types of adhesives
UV and light-curable adhesives
Temporary masking resins
Conformal coatings
Potting materials
Encapsulants
FIP/CIP- gaskets
In addition: Materials with secondary moisture- and heat cure options, adhesive activators.

For applications in the field of
Medical
Orthopedic implants
Electronics
Automotive
Aerospace
Glass
Optical Bonding

Additional Products
UV-Spot and flood lamps (Broadband and LED)
UV-Conveyors
Radiometers
Dispensing equipment
Technical consulting
System Integration

ekp coatings GmbH
Ersteiner Straße 11
D-79346 Endingen am Kaiserstuhl
Phone +49 (0) 7642-9260-60
Fax +49 (0) 7642-9260-500
Email: info@ekp-coatings.com
www.ekp-coatings.com

Member of IVK

Company

Year of formation
2000

Size of workforce
50 employees

Nominal capital
120,000 Euro

Ownership structure
owner Stefan Ermisch

Sales channels
directly

Contact partners
Management:
Stefan Ermisch
Email: stefan.ermisch@ekp-coatings.com

Application technology and sales:
Technical product management/technical
customer service Darja Rebernik
Email: darja.rebernik@ekp-coatings.com

Further information
Expert for Packaging Solutions: development
and production of special and functional
coatings for Flexible Packaging, Rigid Metal
Packaging and Pharmaceuticals, technical
know how in combination with extensive
consulting and individual customer support.

Range of Products

Types of adhesives
Solvent-based adhesives
Dispersion adhesives

For applications in the field of
Paper/packaging
Bookbinding/graphic design
Adhesive tapes, labels
Household, recreation and office

ADHESIVES • COATINGS

Eluid Adhesive GmbH
Heinrich-Hertz-Straße 10
D-27283 Verden
Phone +49 (0) 42 31-3 03 40-0
Fax +49 (0) 42 31-3 03 40-17
Email: info@eluid.de
www.eluid.de

Member of IVK

Company

Year of formation
1932

Size of workforce
7

Managing partners
Andreas May

Ownership structure
100 % Private

Contact partners
Andreas May
Karin Münker

Sales channels
Europe: through our own sales force,
traders and agents
Worldwide: traders and agencies

Range of Products

Types of adhesives
Acrylate, Styrene-Acrylate, Polyurethane-
and Vinyl acetate dispersions
(A, RA, SA, PUD, PVAC, VAE) for adhesives,
varnish and coatings
PSA Dispersions
PVOH adhesives
Dextrin-, Casein- and starch adhesives.
APAO, EVA, PSA, PO, PUR Hotmelts

For applications in the field of
Coatings for films (printable)
Coatings for films (Soft Touch)
Bookbinding/graphic arts
Paper and Converting
Folding boxes
Packaging
Envelope industry
Protection of books
Adhesives for insulation technology
Tapes (single and double sided) / Labels
Flexible packaging
High-gloss film laminating
Pressure sensitive adhesives for film
(removable/permanent)
Heat Seal Products
Wood adhesives D2 / D3
Foam processing industry
Protection foil
Safety documents
Nonwoven industry
Wallpaper industry
Textile industry
Transformerboards

Emerell AG
Neulandstrasse 3
6203 Sempach Station
Switzerland
Phone +41 41 469 91 00
Email: info@emerell.com
www.emerell.com

Member of FKS

Company

Emerell is specialized in the contract manufacturing of chemical-technical products. Our core competencies and decades of experience lie in mixing technology, polymerization and scale-up. Emerell's services are aimed at large companies and SMEs that wish to outsource part of their production, as well as start-ups that do not yet have their own production facilities. Emerell concentrates on pure contract manufacturing and enjoys a high level of trust by not using its own products. From the pilot phase to series production, Emerell is your partner for customized solutions and highest industrial demands as a pure contract manufacturer.

Owner
Adrian Leumann

Contact
Dr. Michael Lang, CEO
Phone +41 41 469 93 14,
E-Mail: michael.lang@emerell.com

Range of Products

Technologies
Emulsionspolymerization
Anionic/Cationic

Solutionpolymerization

Mixing Technologies for liquid, paste
1-component PUR/2-component PUR
adhesives
Dispersions
UV-crosslinkable adhesives
Custom mixes (liquid and paste-like
products)

Others
Aqueous adhesives
Micro-Beads
Microencapsulation

Industries
Construction and building industry
Electronics
Labels, adhesive tapes, packaging
Optics
System providers
Shipping industry
General paint- and coating industry
Adhesives
Others

EMS-CHEMIE AG
Business Unit EMS-GRILTECH
Via Innovativa 1
7013 Domat/Ems
Phone +41 81 632 72 02
Fax +41 81 632 74 02
Email: info@emsgriltech.com
www.emsgriltech.com

Member of FKS

Company

Year of formation
1936 founded as Holzverzuckerungs AG (HOVAG), 1960 renamed into EMSER WERKE AG and finally into EMS-CHEMIE AG in 1981.

Size of workforce
2,521 worldwide in December 2020

Sales channels
Direct Sales Channel, Distributors/Traders

Contact partners
Application technology and sales:
Phone: +41 81 632 72 02, Fax: +41 81 632 74 02
Email: info@emsgriltech.com, www.emsgriltech.com

Contact partners
The business unit EMS-GRILTECH is part of EMS-CHE-MIE AG which belongs the EMS-CHEMIE HOLDING AG. We manufacture and sell Grilon, Nexylon and Nexylene fibers, Griltex hotmelt adhesives, Grilbond adhesion promoters, Primid crosslinkers for powder-coatings and Grilonit reactive diluents. We have developed these materials and additives into excellent specialty products for technically demanding applications. In this way we create added value for our customers supporting their continual improvements.

Thermoplastic hotmelt adhesives
Thermoplastic adhesive products for technical and textile bonding applications are sold under the trade name „Griltex®". EMS-GRILTECH has many years of experience in the manufacture of tailor-made copolyamides and copolyesters for different application fields. The melt temperatures and melt viscosity can be modified over a wide range depending on the different requirements of each application. The adhesives are available as powder in a wide range of grain sizes or as granules. Manufacturing is carried out in our own polymerisation and grinding plants.

Griltex® ES – Bonding of Even Surfaces
Hotmelt adhesives for bonding of metal, plastics, glass and other smooth surfaces are produced under the trade name Griltex® ES.

Range of Products

Types of adhesives
Hot melt adhesives

Types of sealants
Other

Raw materials
Additives
Resins
Polymers

Equipment, plant and components
for conveying, mixing, metering and for adhesive application
measuring and testing

For applications in the field of
Paper/packaging
Wood/furniture industry
Construction industry, including floors, walls and ceilings
Electronics
Mechanical engineering and equipment construction
Automotive industry, aviation industry
Textile industry
Adhesive tapes, labels
Hygiene
Household, recreation and office

EMS-GRILTECH's corporate offices with research laboratory, technical service centre and production facility are located at Domat/Ems, Switzerland. We also have state-of-the-art production locations, application development centers and customer service laboratories in USA, China, Taiwan and Japan.

EMS-GRILTECH is present worldwide either with its own sales companies or represented by agents.

EUKALIN Spezial-Klebstoff Fabrik GmbH
Ernst-Abbe-Straße 10
D-52249 Eschweiler
Phone +49 (0) 24 03-64 50-0
Fax +49 (0) 24 03-64 50-26
Email: eukalin@eukalin.de
www.eukalin.com

Member of IVK

Company

Year of formation
1904

Size of workforce
70 employees

Managing partners
Timm Koepchen
Jan Schulz-Wachler

Ownership structure
family owned

Subsidiaries
EUKALIN Corp. USA

Sales channels
Directly and through dealers

Contact partners
Application technology and sales:
Timm Koepchen

Range of Products

Types of adhesives
Dispersion adhesives
Pressure-sensitive adhesives
• UVC-curable acrylic hot melts
• Ruber-based hot melts
• Dispersions
Dextrin and starch adhesives
Glutine glue
Hot meld adhesives
Latex adhesives
Cold seal adhesives
Polyurethane Adhesives

Equipment, plant and components
for conveying, mixing, and metering
for adhesive application

For applications in the field of
Paper/packaging
Bookbinding/graphic design
Textile industry
Flexible packaging
Construction industry
Tapes + Labels
Labelling
E-Commerce
Paper straws
Barrier papers

Evonik Industries AG

D-45764 Marl, www.evonik.com/crosslinkers,
www.evonik.com/adhesives-sealants,
www.evonik.com/designed-polymers
D-45764 Marl, www.vestamelt.de
D-64293 Darmstadt, www.visiomer.com
D-45127 Essen, www.evonik.com/polymer-dispersions
www.evonik.com/hanse, www.evonik.com/tegopac
D-63457 Hanau, www.aerosil.com,
www.dynasylan.com, www.evonik.com/fp

Member of IVK

Company

Year of formation
2007

Contact partners
Email: adhesives@evonik.com
Website: adhesives-sealants.evonik.com/en

Company information
Evonik is one of the world leaders in specialty chemicals. The company is active in more than 100
countries around the world and generated sales of
€ 12.2 billion and an operating profit (adjusted
EBITDA) of € 1.91 billion in 2020. Evonik goes far
beyond chemistry to create innovative, profitable
and sustainable solutions for customers. More than
33,000 employees work together for a common
purpose: We want to improve life today and tomorrow.

Range of Products

Types of adhesives
Hot melt adhesives (VESTAMELT®) (DYNACOLL® S)

Types of sealants
Acrylic sealants (DEGACRYL®)

Raw materials
Additives: waxes (VESTOWAX®, SARAWAX®),
defoamer (TEGO® Antifoam), wetting agents
(TEGOPREN®), thickener (TEGO® Rheo), silica na-
noparticles (Nanopox®), silicone rubber particles
(Albidur®), methacrylate monomers (VISIOMER®),
pyrogenic silicas and metal oxides (AEROSIL®,
AEROXIDE®), specialty precipitated silicas
(SIPERNAT®), functional silanes (Dynasylan®)

Crosslinkers: speciality resins, aliphatic diamines
(VESTAMIN®), aliphatic isocyanates (VESTANAT®)

Polymers: amorphous poly-alpha-olefines
(VESTOPLAST®), copolyesters (DYNACOLL®),
liquid polybutadienes (POLYVEST®), polyacrylates
(DEGACRYL®, DYNACOLL® AC), silane-modifiied
polymers (Polymer ST, TEGOPAC®), condensation
curing silicones (Polymer OH)

Epoxy Curing Agents: Polycarbamide curtives for
HDI trimer (Amicure®), Polyamide and amidoamine
curing agents (Ancamide®), Aliphatic & cycloali-
phatic amine curing agents (Ancamine®)

For applications in the field of
Paper/packaging
Bookbinding/graphic design
Wood/furniture industry
Construction industry, including floors,
Walls and ceilings
Electronics
Automotive industry, aviation industry
Textile industry
Adhesive tapes, labels
Hygiene
Manufacturing of hotmelt adhesives
Wind energy

Fenos AG

Steinheimer Straße 3
D-71691 Freiberg a.N.
Phone +49 (0) 7141 992249-0
Fax +49 (0) 7141 992249-99
Email: info@fenos.de
www.fenos.de

Member of IVK

Company

Year of formation
2015

Size of workforce
10

Sales channels
Technical Sales and representatives
international

Contact partners
Application technology and sales:
Dr. Natalia Fedicheva

Further information
Fenos AG is an independant system house
for industrial adhesives and PUR formulations.
All products are developed towards
customers' requirements and processes.

Range of Products

Types of adhesives
Hot melt adhesives
Reactive adhesives
Pressure-sensitive adhesives

For applications in the field of
Paper/packaging
Wood/furniture industry
Construction industry, including floors,
walls and ceilings
Automotive industry, aviation industry
Textile industry
Adhesive tapes, labels
Hygiene
Household, recreation and office

Fermit GmbH

Zur Heide 4
D-53560 Vettelschoß
Phone +49 (0) 26 45-22 07
Fax +49 (0) 26 45-31 13
Email: info@fermit.de
www.fermit.com

Member of IVK

Company

Year of formation
2008

Size of workforce
15

Ownership structure
100% subsidiary of Barthélémy/France

Sales channels
sanitary and heating equipm. trade,
industry, whole trade

Contact partners
Management:
Alois Hauk

Further information
former Nissen & Volk, Hamburg

Range of Products

Types of adhesives
Solvent-based adhesives
Resin-based adhesives

Types of sealants
Silicone sealants
MS/SMP sealants
Sealing pastes
Other

For applications in the field of
Sanitary and heating industry
Construction Industry
Mechanical engineering and equipment
construction
Automotive industry, aviation industry

fischerwerke GmbH & Co. KG

Klaus-Fischer-Straße 1
D-72178 Waldachtal
Phone +49 (0) 74 43 12-0
Email: info@fischer.de
www.fischer.de

Member of IVK

Company

Year of formation
1948

Size of workforce
5,200

Ownership structure
Privately owned

Subsidiaries
50 (AE, AR, AT, BE, BR, BG, CN, CZ, DK, DE, ES, FI, FR, GR, HR, HU, IN, IT, JP, KR, MX, NL, NO, PH, PL, QA, PT, RO, RS, RU, SE, SG, SK, TH, TR, US, UK, VN)

Sales channels
DIY, specialized trade

Contact partners
Michael Geiszbühl
Email: Michael.Geiszbuehl@fischer.de

Further information
World market leader in chemical fastening systems

Range of Products

Types of adhesives
Reactive adhesives
Dispersion adhesives
MS / SMP adhesives
UV curing adhesive
Solvent based adhesives

Types of sealants
Acrylic sealants
Silicone sealants
MS/SMP sealants
Polyurethane sealants
Solvent based sealants
Bitumen based sealants

For applications in the field of
Wood/furniture industry
Construction industry, including floors, walls, ceilings and roofs
Window and door construction
Carpenter and stair construction
Metal construction
DIY

Follmann GmbH & Co. KG

Heinrich-Follmann-Straße 1
D-32423 Minden
Phone +49 (0) 5 71-93 39-0
Fax +49 (0) 5 71-93 39- 300
Email: adhesives@follmann.com
www.follmann.com

Member of IVK

Company

Year of formation
1977

Management
Dr. Jörn Küster
Dr. Jörg Seubert

Subsidiaries
OOO Follmann
(Moscow/Russian Federation)
Follmann (Shanghai) Trading Co., Ltd.
(Shanghai/China)
Follmann Chemia Polska (Poznan/Poland)
Sealock Sp. z o.o. (Warsaw/Poland)
Sealock Ltd. (Andover/Great Britain)

Contact partners
Holger Nietschke
Sebastian Drewes
Martin Haupt

Range of Products

Types of adhesives
Hotmelt adhesives
Dispersion adhesives
Reactive hotmelt adhesives
Pressure-sensitive adhesives
Starch and casein adhesives

For applications in the field of
Paper/packaging
Bookbinding/graphic design
Wood + Furniture industry
Automotive industry
Assembly
Textile industry
Adhesive tapes
Labeling
Upholstery, mattresses
Filtration industry
Abrasive industry
Flat Lamination
Sandwichpanels, Caravan
Transportation
Building & construction
Flooring

H.B. Fuller

Connecting what matters.™

H.B. Fuller Europe GmbH
Talacker 50
CH-8001 Zürich
www.hbfuller.com

Member of IVK, FKS, VLK

Company

Global reach
- H.B. Fuller operates three regional headquarters across the globe:
  Americas – St. Paul, Minn., U.S.
  EIMEA – Zurich, Switzerland
  Asia Pacific – Shanghai, China
- Direct presence in 36 countries and customers in more than 125 geographic markets.

European commitment
H.B. Fuller has a network of specialised production sites across Europe serving customers in electronics, disposable hygiene, medical, transportation, aerospace, clean energy, packaging, construction, woodworking, general industries and other consumer businesses.

About H.B. Fuller
Since 1887, H.B. Fuller has been a leading global adhesives provider focusing on perfecting adhesives, sealants and other specialty chemical products to improve products and lives. With fiscal 2020 net revenue of over $ 2.8 billion, H.B. Fuller's commitment to innovation brings together people, products and processes and sustainability that answer and solve some of the world's biggest challenges. Our reliable, responsive service creates lasting, rewarding connections with customers for a better tomorrow. And, our promise to our people connects them with opportunities to innovate and thrive. For more information, visit us at www.hbfuller.com.

Range of Products

Our Technologies
- Hot Melt
- Polymer and Specialty Technologies
- Reactive Chemistries: Urethane Epoxy Solventless
- Solvent-based
- Water-based

Our markets
- Automotive
- Building and Construction
- Consumer Products
- Electronic and Assembly Materials
- Emulsion Polymers
- General Assembly
- Nonwovens and Hygiene
- Packaging
- Paper Converting
- Woodworking

Committed to our communities
- Supporting STEM education and youth leadership development
- Volunteers reach more than 30 countries with 6,600-plus hours of service annually
- Company and employee donations total more than $ 1.2 million globally each year.

GLUETEC GROUP

GLUETEC GROUP –
GLUETEC Industrieklebstoffe GmbH & Co. KG
Am Biotop 8a
D-97259 Greußenheim
Phone +49 (0) 9369 9836-0
Fax +49 (0) 9369 9836-10
Email: info@gluetec-group.com
www.gluetec-group.com

Company

Year of formation
1997

Size of workforce
75

Managing partners
Christine Kopp, Willi Kopp

Ownership structure
Family business

Subsidiaries
GLUETEC Germany:
GLUETEC Industrieklebstoffe GmbH & Co. KG
WIKO Poland: WIKO Klebetechnik Sp. zo.o.
GLUETEC Slovenia: GLUETEC d.o.o.

Sales channels
Sales force, subsidiaries abroad, dealers,
private label business, contract filling

Contact partners
Management: Christine Kopp
Sales management: Nils Lang
Product management: Dr. Benedikt Peter

Further information
GLUETEC GROUP is the right partner for integrated adhesive solutions in the B2B environment.
GLUETEC has more than 20 years of experience
with chemical products for bonding & sealing for
industry, craft and trade. With the courage to
break new ground, GLUETEC develops and produces its own high-performance formulations.
As a full-service provider, GLUETEC scores with
attractive services along the value chain and
enjoys the trust of renowned customers in all
areas of industry and trade. The basis of all actions
is the thinking in holistic adhesive solutions. With
our adhesives expertise, we support you in all
aspects of product selection, packaging, product
design, private label, contract filling, contract
labelling, packaging and shipping. GLUETEC
scores with a large product portfolio that was
developed specifically for customer applications.
Tailor-made products are successfully marketed

Range of Products

Types of adhesives
Reactive adhesives
Solvent-based adhesives
Spray adhesives
1-component base: cyanacrylat,
polyurethan, anaerobic adhesives, PUR
2-component-adhesives base: epoxy resin, PUR,
methyl meth acrylate

Types of sealants
PUR sealants; Silicone sealants; MS/SMP sealants

Raw materials
Fillers; Resins; Polymers

Equipment, plant and components
For conveying, mixing, metering and for adhesive
application

For applications in the field of
Graphic design
Wood/furniture industry
Construction industry, including floors,
walls and ceilings
Electronics
Mechanical engineering and equipment
construction
Automotive industry, automotive after market,
aviation industry, textile industry, railway industry
Adhesive tapes, labels
Hygiene
DIY

as own brands and as individual private label
solutions. A large selection of innovative
containers, cartridges and syringes always
guarantees the best possible handling of the
products in manual bonding. GLUETEC's products
are used in a wide variety of areas - from
production and maintenance to repair.

GLUDAN
Deutschland GmbH

Am Hesterkamp 2
D-21514 Büchen
Phone +49 (0) 41 55-49 75-0
Fax +49 (0) 41 55-49 75 49
Email: gludan@gludan.de
www.gludan.com

Member of IVK

Company

Year of formation
1977

Size of workforce
28

Ownership structure
GmbH

Sales channels
Own sale force, and agents, have a look on our map www.gludan.com for contacts

Contact partners
Management:
ks@gludan.de

Application technology and sales:
od@gludan.de
sb@gludan.de

Further information
www.gludan.com

Range of Products

Types of adhesives
Hot melt adhesives
Dispersion adhesives
Glutine glue
Pressure-sensitive adhesives

Raw materials
Additives
Fillers
Polymers
Starch

Equipment, plant and components
for conveying, mixing, metering and
for adhesive application
measuring and testing

For applications in the field of
Paper/packaging
Bookbinding/graphic design
Wood/furniture industry
Construction industry, including floors, walls and ceilings
Textile industry
Adhesive tapes, labels
Hygiene
Household, recreation and office

Gößl + Pfaff GmbH

Münchener Straße 13
D-85123 Karlskron
Phone +49 8450-9320
Fax +49 8450-932 13
Email: info@goessel-pfaff.de
www.goessl-pfaff.de

Member of IVK

Company

Year of formation
1984

Size of workforce
16

Managing partners
Roland Gößl
Josef Pfaff

Sales channels
Technischer Vertrieb/Web-Shop

Contact partners
Management:
Roland Gößl

Application technology and sales:
Martina Reithmeier

Range of Products

Types of adhesives
Reactive adhesives

Equipment, plant and components
for conveying, mixing, metering and for
adhesive application

For applications in the field of
Wood/furniture industry
Construction industry, including floors, walls
and ceilings
Electronics
Mechanical engineering and equipment
construction
Automotive industry, aviation industry

Grünig KG
Häuserschlag 8
97688 Bad Kissingen
Phone +49 (0) 9736 7571-0
Fax +49 (0) 9736 7571-29
Email: info@gruenig-net.de
www.gruenig-net.de

Member of IVK

Company

Year of formation
1961

Size of workforce
35

Managing partners
Thomas Ulsamer
Sabine Ulsamer

Sales channels
Phone, FAX, Email, Pos

Contact partners
Management:
Thomas Ulsamer
Sabine Ulsamer

Application technology and sales:
Dietmar Itt, Andreas Schwab,
Reinhold Teufel, Michał Kozieł,
Iwona Greczanik

Range of Products

Types of adhesives
Dispersion adhesives
Dextrin

Equipment, plant and components
Polymerization, formulation

For applications in the field of
Paper/Packaging
Tubewinding
Edge protectors
Honey combs
Displays
Graphic design
Plasterboard production
Wood application
Furniture
Flooring

Kompetenz | Qualität | Partnerschaft

GYSO AG
Steinackerstrasse 34
CH-8302 Kloten
Phone +41 43 255 55 55
Email: info@gyso.ch
www.gyso.ch

Member of IVK

Company

Year of formation
1957

Size of workforce
135

Ownership structure
Owner managed family business

Sales channels
Direct to the enduser as well as via specialized trade

Contact partners
Management:
Roland Gysel

Application technology and sales:
Kandid Voegele (Technology),
Thomas Emler (Sales)

Further information
GYSO AG is a Swiss family company founded in 1957. From the very beginning, the company was engaged in adhesives and sealants. In the course of time, sealing tapes, adhesive tapes, films and other product lines have been added.Today, GYSO has a wide and comprehensive range of products, focused on bonding, sealing, protecting, sanding, painting and finishing. The development is always guided by the idea of offering high quality and prac- tical solutions.Our large and loyal clientele from the building industry and the automotive is for us con- firmation and at the same time motivation for the future. From a one-man business, GYSO AG has developed into an efficient and modern company with over 130 employees. A competent partner where customer satisfaction is always in the fore- ground. GYSO AG employs approx. 130 employees, of which over 30 work in the field to serve our customers in all parts of the country and in all language regions. Since the company was founded, the market in the construction, automotive and industrial sectors has changed in unexpected ways. GYSO AG is proud of the fact that it has been able to keep pace with developments over all these years and has always been able to adapt its pro-

Range of Products

Types of adhesives
Hot melt adhesives; Solvent-based adhesives
Dispersion adhesives; Glutine glue

Types of sealants
Acrylic sealants; Butyl sealants; PUR sealants;
Silicone sealants; MS/SMP sealants; Other

Equipment, plant and components
for conveying, mixing, metering and for adhesive application; for surface pretreatment

For applications in the field of
Paper/packaging
Bookbinding/graphic design
Wood/furniture industry
Construction industry, including floors, walls and ceilings
Mechanical engineering and equipment construction
Automotive industry, aviation industry
Textile industry
Household, recreation and office

ducts to the new, sometimes very demanding requirements.Our clearly structured sales orga- nization with dynamic and innovative employees guarantees quick, competent and practical advice. Thanks to many years of experience, combined with profound technical knowledge, we offer individual solutions to problems, adapted to the customer's needs. Our field service team, repre- sented in all regions of Switzerland, is available at any location within a short time. Personal and professional support during planning and execu- tion on site. Factual information and training in the form of specialist conferences at customers' premises and at schools are our contribution to obtaining and maintaining optimum construction quality.

Fritz Häcker GmbH + Co. KG

Im Holzgarten 18
D-71665 Vaihingen/Enz
Phone +49 (0) 70 42-94 62-0
Fax +49 (0) 70 42-9 89 05
Email: info@haecker-gel.de
www.haecker-gel.de

Member of IVK

Company

Year of formation
1885

Size of workforce
22

Ownership structure
private owned company

Sales channels
Distributers worldwide

Contact partners
Management:
Ralf Müller

Range of Products

Types of adhesives
Gelatine based Adhesives

Types of sealants
Other

Raw materials
Technical Gelatine

Equipment, plant and components
for conveying, mixing, metering and for
adhesive application
measuring and testing

For applications in the field of
Paper/packaging
Bookbinding/graphic art
Tissue and towels
Box covering and lamination

HANSETACK GmbH

Saseler Straße 182
D-22159 Hamburg
Phone +49 (0) 40-237 242 67
Fax +49 (0) 40-237 242 69
Email: info@hansetack.com
www.dercol.pt

Member of IVK

Company

Range of Products

Year of formation
2019

Size of workforce
3

Managing partners
Lars-Olaf Jessen

Nominal capital
350,000 €

Ownership structure
private

Contact partners
Management:
Lars-Olaf Jessen

Application technology and sales:
Lars-Olaf Jessen

Raw materials
Resins

For applications in the field of
Paper/packaging
Bookbinding/graphic design
Wood/furniture industry
Construction industry, including floors,
walls and ceilings
Electronics
Automotive industry, aviation industry
Textile industry
Adhesive tapes, labels
Hygiene
Household, recreation and office

Henkel
AG & Co. KGaA
Henkelstraße 67
D-40191 Düsseldorf
Phone +49 (0) 211-797-0
www.henkel.com

Member of IVK, FCIO, FKS, VLK

Company

Ownership structure
AG & Co. KGaA

Contact partners
Business Unit Adhesive Technologies
Headquarters Düsseldorf:
Phone +49 (0) 211-797-0
www.henkel-adhesives.com

Henkel Central
Eastern Europe GmbH
Phone +43 (1) 7 11 04-0
www.henkel.at

Henkel & Cie. AG
Phone +41 (61) 825-70 00
www.henkel.ch

Henkel Belgium N.V.
Phone +32 (2) 421-27 11
www.henkel.be

Further information
Henkel Adhesive Technologies is leading today's markets and shaping tomorrow's through its adhesives, sealants and functional coatings. We enable the transformation of entire industries, giving our customers a competitive advantage and offering consumers a unique experience. Innovative thinking and entrepreneurial spirit are part of our DNA. Being industry and application experts across manufacturing industries worldwide, we work closely with our

Range of Products

Types of adhesives
Hot melt adhesives
Reactive adhesives
Solvent-based adhesives
Dispersion adhesives
Adhesives based on renewable resources
Pressure-sensitive adhesives
Instant adhesives
Structural adhesives
Anaerobic Adhesives
Laminating Adhesives

Types of sealants
Acrylic sealants
Butyl sealants
PUR sealants
Silicone sealants
MS/SMP sealants
Other

customers and partners to create sustainable value for all stakeholders – with trusted brands and high-impact solutions based on an unmatched technology portfolio.

Our industrial product portfolio is organized into five Technology Cluster Brands - Loctite, Technomelt, Bonderite, Teroson and Aquence. For consumers and craftsmen, we focus on the four global brand platforms Pritt, Loctite, Ceresit and Pattex.

DRIVING PROGRESS TOWARDS A CIRCULAR ECONOMY

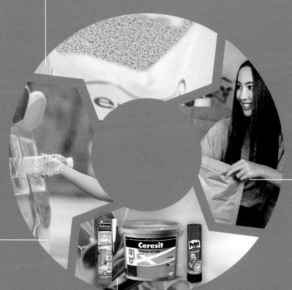

Raw Materials
Use of renewable raw materials

Design
Technologies enabling the replacement of plastics, e.g. in padded mailers

Recycling
Technologies for recycling compatible labeling, e.g. for PET bottles

Production
Launch of recycled materials for product packaging

We are leading with technologies and solutions for adhesives, sealants and functional coatings and focus our capabilities on the development of innovations that help contributing to solve ecological and societal challenges today and tomorrow. Based on our innovative portfolio we improve energy efficiency, material use, the durability and recyclability of a broad variety of products across the entire value chain and thereby drive progress towards a circular economy.

IMCD N.V.
Wilhelminaplein 32
3072 DE Rotterdam, Netherlands
Phone +31 10 290 86 84
Fax +31 10 290 86 80
Email: coatings.construction@imcdgroup.com
www.imcdgroup.com/business-groups/
coatings-construction

Member of IVK

Company

Ownership structure
IMCD N.V. is a leading company in sales, marketing and distribution of specialty chemicals and food ingredients. With a revenue of € 2,379 million and 2,800 professionals in more than 50 countries, IMCD provides its partners with optimum tailored solutions for multi-territory distribution management in EMEA, Asia-Pacific and Americas.

Management
Piet van der Slikke, CEO IMCD N.V.
Frank Schneider, Director Business Group Coatings & Construction

Contact partners
Germany & International
Heinz J. Küppers, Market Manager Adhesives
Email: heinz.kueppers@imcd.de

Switzerland:
Armin Landolt, Sales Manager Adhesives
Email: armin.landolt@imcd.ch

Austria and South East Europe:
Szabolcs Antal, Business Unit Manager
SEE - Coatings & Construction, Synthesis, Lubricants & Fuels
Email: szabolcs.antal@imcd.at

Scandinavia:
Xoan Au, Business Unit Manager Coatings & Construction
Email: xoan.au@imcd.se

Range of Products

Types of adhesives
Solvent-based adhesives
Solvent-free adhesives
Dispersion adhesives
Pressure sensitive adhesives

Raw materials
Additives, Functional Fillers, Resins, Pigments, Solvents, Polymers

For applications in the field of
Paper/Packaging
Woodworking/Joinery
Building, Construction, Civil Engineering
Electronics
Automotive
Textile
Tapes, Labels, Graphic Arts

Services
Customer support
Guide formulations
Seminars

Jobachem GmbH
Am Burgberg 13
D-37586 Dassel
Phone +49 (0) 5564-20078-0
Fax +49 (0) 5564-20078 - 11
Email: info@jobachem.com
www.jobachem.com

Company

Year of formation
1992

Size of workforce
50 employees worldwide

Managing partners
Julian Kahl

Nominal capital
1.000.000 €

Ownership structure
Family business

Subsidiaries
JOBACHEM Suzhou Co., Ltd.; JOBACHEM L.P.

Sales channels
B2B

Contact partners
Management:
Dr. Ivonne Anita Ehli

Application technology and sales:
Dr. Ivonne Anita Ehli

Range of Products

Types of adhesives
Hot melt adhesives
Reactive adhesives
Solvent-based adhesives
Dispersion adhesives
Glutine glue
Pressure-sensitive adhesives

Types of sealants
Acrylic sealants
Butyl sealants
PUR sealants
Silicone sealants
Other

Raw materials
Additives
Resins
Solvents

Equipment, plant and components
for conveying, mixing, metering and for adhesive application

For applications in the field of
Paper/packaging
Bookbinding/graphic design
Wood/furniture industry
Construction industry, including floors, walls and ceilings
Electronics
Mechanical engineering and equipment construction
Automotive industry, aviation industry
Textile industry
Adhesive tapes, labels
Hygiene
Household, recreation and office

Klebstoffe

Jowat SE
Ernst-Hilker-Straße 10 – 14
D-32758 Detmold
Phone +49 (0) 52 31-7 49-0
Email: info@jowat.de
www.jowat.com

Member of IVK, FKS

Company

Year of formation
1919

Size of workforce
approx. 1,200

Managing partners
Klaus Kullmann
Ralf Nitschke
Dr. Christian Terfloth

Ownership structure
Shareholder company (not publicly traded)

Subsidiaries
23 worldwide

Sales channels
Own affiliations and distributors

Contact partners
Productmanagement
Ingo Horsthemke

Range of Products

Types of adhesives
Hot melt adhesives
Reactive adhesives
Solvent-based adhesives
Dispersion adhesives
Pressure-sensitive adhesives
Separating agents

For applications in the field of
Paper/packaging
Bookbinding/graphic design
Wood/furniture industry
Construction industry, including floors, walls and ceilings
Electronics
Automotive industry, aviation industry
Textile industry
Adhesive tapes, labels
Upholstery, mattresses

Kaneka

KANEKA BELGIUM NV

The Dreamology Company
—Make your dreams come true—

Kaneka Belgium N. V.
MS Polymer Division
Nijverheidsstraat 16
B-2260 Westerlo (Oevel)
E-Mail: info.mspolymer@kaneka.be
www.kaneka.be

Member of IVK

Company

Year of formation
Kaneka Belgium N.V. was founded in 1970 as the European production site of the globally acting Kaneka Corporation, Japan.

Further information
Kaneka products have conquered the European market, becoming a synonym for premium quality raw materials, with the brand MS POLYMER basically defining a new group of adhesives and sealants. Other brands of Kaneka's MS Polymer Division are SILYL and XMAP.

Production/Technical Service contact
Kaneka Belgium N.V.
MS Polymer Division
Nijverheidsstraat 16
B-2260 Westerlo (Oevel)
Phone +32 14-25 78 67
Email: info.mspolymer@kaneka.de

Marketing contact
Kaneka Belgium N.V.
MS Polymer Division
Nijverheidsstraat 16
B-2260 Westerlo (Oevel)
Phone +32 14-25 45 20
Email: info.mspolymer@kaneka.de

For contact in D, A, CH, CEE
Werner Hollbeck GmbH
Karl-Legien-Straße 7
D-45356 Essen
Phone +49 (0) 2 01-7 22 16 16
Email: info@hollbeck.de

Range of Products

Raw materials
MS Polymer, SILYL, and XMAP are reactive high performance polymers based on polyether, or polyacrylate, respectively. Customers can select between moisture curing grades and radical cure types including UV-cure. The polymers' characteristics produce elastic adhesives and sealants, but also high strength types and coatings can be formulated.

End products' features
Solvent and isocyanate free 1-p/2-p-reactive adhesives and sealants
Pressure sensitive adhesives (PSA)
Oil resistance, temperature resistance (150 °C permanent use, XMAP)
Low gas and moisture permeability
High UV-resistance
Adhesive blends with epoxies
MS-Emulsions

For applications in the field of
Wood/furniture industry
Construction industry, including floors, walls and ceilings
Electronics
Automotive industry, transportation industry
Adhesive tapes, labels
Household, recreation and office
General industry
Shipbuilding industry
Waterproofing and roof coatings

KEYSER & MACKAY

Keyser & Mackay
German branch office
Ettore-Bugatti-Straße 6 –14
D-51149 Köln
Phone +49 (0) 2203-20301-0
Fax +49 (0) 2203-20301-33
Email: keymac.de@keymac.com
www.keysermackay.com

Member of IVK

Company

Founded
1894

Size of workforce
120

Subsidiaries
Headquarter in The Netherlands

Subsidiaries in Germany, Belgium, France, Switzerland, Poland, Spain

Contact partners
Mr. Robert Woizenko
Email: r.woizenko@keymac.com

Types of adhesives
Hot melt adhesives
Reactive adhesives
Solvent-based adhesives
Dispersion adhesives
Pressure-sensitive adhesives

Types of sealants
Acrylic sealants
Butyl sealants
Polysulfide sealants
PUR sealants
Silicone sealants
MS/SMP sealants

Range of Products

Raw materials
Resins: hydrogenated hydrocarbon resins, C5/C9 hydrocarbon resins, pure monomer resins, modified rosin resins, resin dispersions

Polymers: amorphous polyolefins (APO), EVA, SIS/ SBS/ SEBS, PE / PP waxes, FT waxes, STP, acrylic copolymers, acrylic dispersions

Fillers: precipitated calcium carbonate, talc, dolomite

Additives: antioxidants, UV absorbers, silanes, oxazolidines, adhesion promoters, flame retardants, polyols

For application in the field of
Paper/packaging
Bookbinding/graphic design
Wood/furniture industry
Construction industry, including floors, walls and ceilings
Electronics
Mechanical engineering and equipment construction
Automotive industry, aviation industry
Textile industry
Adhesive tapes, labels
Hygiene
Household, recreation and office

Kiesel Bauchemie GmbH u. Co. KG

Wolf-Hirth-Straße 2
D-73730 Esslingen
Phone +49 711 93134-0
Fax +49 711 93134-140
Email: kiesel@kiesel.com
www.kiesel.com

Member of GEV

Company

Year of formation
1959

Size of workforce
160

Ownership structure
family owned

Subsidiaries
Sales offices in Benelux, Poland, Czech Republic, Switzerland, France

Sales channels
wholesale to professional installers

Contact partners
Management & Sales:
Beatrice Kiesel-Luik

Head of Export and Marketing:
Alexander Magg

Range of Products

Types of adhesives
Reactive resin adhesives
SMP and Silane modified
SMP adhesives-Products
Dispersion adhesives
Cement based adhesives

Types of sealants
Acrylic sealants
PUR sealants
Cement based sealants
Dispersion sealants

For applications in the field of
Construction sites/industry, including floors, walls and ceilings
Inside and Outside

Kisling

Kisling AG
Motorenstrasse 102
CH-Wetzikon
Email: info@kisling.com
Tel +41 58 272 0101
Member of IVK, FKS

Company

Year of formation
1862

Size of workforce
> 50

Sales channels
Direct sales, distributors

Contact partners
Managing Director:
Dr. Dirk Clemens

Chief Sales Officer:
Roger Affeltranger

Range of Products

Types of adhesives
Methacrylate Structural Adhesives
No-Mix Structural Adhesives
Anaerobic Adhesives
Cyanoacrylate Adhesives
RTV Silicones
Hybrid Polymers
Structural Epoxy Adhesives

For applications in the field of
Electronics
Automotive
Maintenance, Repair & Overhaul
Lightweight/Composites
E-Mobility/E-Motors
Fluid Technology
Transportation
Loudspeaker

KLEIBERIT Adhesives
KLEBCHEMIE M.G. Becker GmbH & Co. KG
Max-Becker-Straße 4
76356 Weingarten/Germany
Phone +49 7244 62-0
Fax +49 7244 700-0
Email: info@kleiberit.com
www.kleiberit.com

Member of IVK

Company

Year of formation
1948

Size of workforce
over 600 worldwide

Ownership structure
Owner Managed GmbH & Co. KG

Subsidiaries
Australia, France, USA, Canada, UK, Japan,
China, Singapore, Russia, Brazil, India,
Mexico, Ukraine, Turkey, Belarus

Sales channels
Direct, Wholesale

Contact partners
Management:
Dipl. Phys. Klaus Becker-Weimann
Leonhard Ritzhaupt

Sales:
Leonhard Ritzhaupt

Further information
Specialist in PUR-adhesive-technology
and surface finishing
Competence PUR

Range of Products

Types of adhesives
PUR Hotmelt (ME)
Reactive Hotmelt (PUR, POR)
Hotmelt (EVA, PO, PA)
1-C and 2-C reactive adhesives
(PUR, STP, Epoxy)
PUR foam systems
Dispersion adhesives
(Acrylat, EVA, PUR, PVAC)
Sealants and assembly adhesives
Pressure-sensitive adhesive
EPI-systems
Solvent-based adhesives

Coating systems
Kleiberit HotCoating® based on PUR
TopCoating based on UV lacquer

For applications in the field of
Wood- and furniture industry
Profile wrapping
Construction industry including floors, walls,
ceilings and façade elements
Sandwich Panels
Textile industry
Automotive industry
Ship and boat building
Bookbinding industry
Surface finishing
Doors, windows, stairs and flooring
Filtration industry
Paper and packaging industry

Kömmerling Chemische Fabrik GmbH
Zweibrücker Straße 200, D-66954 Pirmasens
Phone +49 (0) 63 31 56-20 00, Fax +49 (0) 63 31 56-19 99
Email: info.koe@hbfuller.com, www.koe-chemie.de
Member of IVK

Company

Year of formation
1897

Size of workforce
450

Ownership structure
H.B. Fuller

Sales channels
B2B, Trading

Further information
Kömmerling Chemische Fabrik GmbH is a leading international manufacturer of high quality adhesives and sealants. Established over 120 years ago, Kömmerling is today a major systems supplier for the Glass, Transport, Construction, Industrial Assembly and Renewable Energy industries. Kömmerling today is part of H.B. Fuller creating a global Top 2 pure play adhesive company.

Range of Products

Types of adhesives
Hot melt adhesives
Reactive adhesives
Solvent-based adhesives
Dispersion adhesives
Pressure-sensitive adhesives

Types of sealants
Butyl sealants
Polysulfide sealants
PUR sealants
Silicone sealants
MS/hybrid sealants

For applications in the field of
Construction
Automotive
Marine
Coil coating
Shoe industry
Insulating glass
Window Bonding
Structural glazing
Wind energy
Renewable energy
Electronics

KRAHN Chemie Deutschland GmbH
Grimm 10
D-20457 Hamburg
Phone +49 (0) 40-3 20 92-0
Fax +49 (0) 40-3 20 92-3 22
Email: info.de@krahn.eu
www.krahn.de

Member of IVK

Company

KRAHN Chemie Deutschland is a speciality chemical distribution company and represents renowned, globally operating manufacturers in Europe. One of our core segments is the adhesives & sealants industry to which we provide raw materials.

Year of formation
1972

Size of workforce
190

Managing directors
Dr. Rolf Kuropka

Ownership structure
Otto Krahn (GmbH & Co.) KG founded 1909

Subsidiaries
KRAHN Chemie Polska Sp. z o.o.
KRAHN Chemie Benelux BV
KRAHN Italia S.p.A.
KRAHN France SAS
InterActive S.A. (Greece)
KRAHN Nordics AB
Pemco-Trigueros Additives Spain S.L.
Perico Ltd. (Great Britain)

Contact partners
Business Segment Manager
Adhesives & Sealants
Thorben Liebrecht
Email: thorben.liebrecht@krahn.eu
Application Technology and Sales:
Marcus Wedemann
Email: marcus.wedemann@krahn.eu

Supplying partners
BASF, Toyal, Valtris, Eastman, Exxon Mobil,
Celanese, Chromaflo, Lanxess, Parker Lord,
OQ Chemicals, Tosoh, Dynaplak, Budenheim

Range of Products

Raw materials
Additives:
Flame retardents, Pigments,
Pigment Pastes,
Rheology Modifiers,
Plasticizers, Biocides

Dispersions:
Acrylic Dispersions
Polychloroprene Latex
Biobased Dispersions

Resins and Hardeners:
Saturated Polyester Resins
EH Diol
Epoxy Hardeners

Polymers:
CR, CSM, EVA, PiB, PVB

Adhesion Promoters

Fillers

For applications in the field of
Paper/packaging
Bookbinding/graphic design
Wood/furniture industry
Construction industry, including floors,
walls and ceilings
Electronics
Traffic Industry, Automotive industry, aviation
industry
Textile industry
Adhesive tapes, labels
Mechanical engineering and equipment
construction DIY

LANXESS Deutschland GmbH

Kennedyplatz 1
D-50569 Cologne
Phone +49 (0) 221 - 88 85- 0
www.lanxess.com

Member of IVK

Company

The global chemical company LANXESS with head office in Cologne, Germany, has been listed at the German stock exchange since 2005. More than 14,300 employees are working for LANXESS worldwide.

The product portfolio consists of 4 segments:
• Advanced Intermediates
• Specialty Additives
• Consumer Protection
• Engineering Materials
The turnover was EUR 6.1bn in 2020.

Range of Products

Biocides
Wide range of biocides under the brand names Preventol® and Metasol®:
• In-can preservatives for all kinds of water-based adhesives
• Fungicides for film protection of sealants, adhesives and grout fillings
• Consultation on microbiological questions

Contact – for biocides
Pietro Rosato
Material Protection Products
Technical Marketing - Industrial Preservation
Phone: +49 (0) 221 8885 7251
Email: pietro.rosato@lanxess.com

L&L Products

L&L Products Europe
Hufelandstraße 7
D-80939 München
Phone +49 (0) 89-413277-990
Email info.europe@llproducts.com
www.llproducts.com

Member of IVK

Company

Year of formation
1958 in Romeo USA –
2002 foundation of the GmbH

Size of workforce
1,200 worldwide

Managing partners
L&L Products Holding Inc.

Nominal capital
25,000 €

Ownership structure
Familys Lane & Ligon

Sales channels
Direct and Distribution

Contact partners
Management:
Matthias Fuchs

Application technology and sales:
Jean-Michel Hollaender

Range of Products

Types of adhesives
Hot melt adhesives
Reactive adhesives
Pressure-sensitive adhesives

Types of sealants
Acrylat sealants
PUR-sealants
MS/SMP sealants
Other

For applications in the field of
Electronics
Mechanical engineering and equipment construction
Automotive industry, aviation industry

Lohmann GmbH & Co. KG
Irlicher Straße 55
D-56567 Neuwied
Phone +49 (0) 26 31-34-0
Fax +49 (0) 26 31-34-66 61
Email: info@lohmann-tapes.com
www.lohmann-tapes.com

Member of IVK

Company

Year of formation
1851

Size of workforce
approx. 1,800 worldwide

Managing directors
Dr. Jörg Pohlman
Dr. Carsten Herzhoff

Subsidiaries
I, F, E, PL, A, GB, NL, DK, SE, RU, UA, USA, China, Korea, Singapore, India, Mexico and Turkey

Sales channels
Market segments:
Graphics, Home Appliances, Industrial, Transportation, Mobile Communication, Medical Technology, Renewable Energies, Electronics and Hygiene

Contact partners
Katharina Kunke,
PR & Communications Manager

Sonja Schöbitz,
PR & Communications Manager

Further information
Lohmann offers mainly customized bonding solutions and takes care of its customers from the first idea up to automatic applications.
This is also shown in the company logo: „The Bonding Engineers".

Range of Products

Types of adhesives
Solvent-free adhesives
Pressure-sensitive adhesives
Reactive adhesives
Dispersion adhesives
Hotmelt adhesives

Types of sealants
Acrylic sealants

Raw materials
Polymers

Equipment, plant and components for
Conveying, mixing, metering
Adhesive application
Surface pretreatment
Adhesive curing and drying
Measuring and testing

For applications in the field of
Paper/packaging
Plate mountig tapes
Wood/furniture industry
Building & Construction industry, including floors, walls and ceilings, structural glazing
Electronics
Automotive industry
Adhesive tapes, die-cuts, labels
Mobile Communication
Renewable Energies
Medical Technology/diagnostics
Hygiene
Home Appliance
High security cards

A clear case for LOHMANN

When it comes to unusual challenges:
we will bond it for you.

Uniting the amazing abilities of structural bonding with the easy handling of regular bonding tape?
Yes, we can do that for you:
Our DuploTEC® SBF (Structural Bonding Films) range offers a powerful alternative to conventional
bonding solutions. It is just one example of how we cover the entire value chain including cutting,
slitting and die-cutting and coating. We provide customer-focused solutions for many industries
including Home Appliances & Electronics, Automotive, Building and Construction.

The Bonding Engineers

www.lohmann-tapes.com

LORD Germany GmbH
Itterpark 8
D-40724 Hilden
Phone +49 (0) 21 03-2 5 23 10
Fax +49 (0) 21 03-2 52 31 97
Email: Liv.Kionka@Parker.com
www.lord.com/emea

Member of IVK

Company

Founded in
1917 by Arthur L. Parker

Employees
56,000 (Parker Hannifin worldwide)

European Technical Center Hilden

Managing Directors
Dr. Dirk L. Schröder, Cornelis Johannes Veraart

Technical Service Contacts in Hilden
LORD Rubber to Metal adhesives
Malte Reppenhagen
Email: Malte.Reppenhagen@Parker.com

LORD Automotive OEM 2K adhesives
Dipl. Ing. Marcus Lämmer
Email: Marcus.Lammer@Parker.com

LORD Industrial adhesives
Dipl. Ing. Marcus Lämmer
Email: Marcus.Lammer@Parker.com

LORD Electronic materials
Christophe Dos Santos
Email: Christophe.DosSantos@Parker.com

LORD Flock adhesives and rubber coatings
Dr. Christiane Stingel
Email: Christiane.Stingel@Parker.com

Additional Information
LORD Corp. is the global leader in producing rubber to metal adhesives used in a wide variety of automotive, aerospace and industrial applications. In addition, LORD is producing solvent based and aqueous flock adhesives and coatings for elastomeric components. On the structural assembly part LORD is producing 2K metal and composite structural adhesives

Range of Products

Type of Adhesives
2K Structural Adhesives (LORD®, FUSOR®, VERSILOK®)
Solvent based rubber to metal adhesives (CHEMLOK®, CHEMOSIL®)
Solvent based reactive 1K adhesives (LORD®, CHEMLOK®, FLOCKSIL®)
Water based reactive 1K adhesives (CUVERTIN®, SIPIOL®)

For applications
Electronics: potting, coating, wafer adhesives
Industrial: metal, composite, plastic
Elastomeric/Rubber: Rubber to metal, flock, slip-coating, anti-squeeze
Automotive (OEM): 2K cold cure hem flange adhesives, 2K repair adhesives, vibration mounts, magnet rheological fluids (MR fluids) and MR devices
Aerospace: 2K OEM structural adhesives, 2K repair adhesives, coatings, vibration mounts, active vibration control devices

which are use e.g. in automotive hem flange bonding. Another focus is on electronic adhesives used for potting and encapsulation as well as electrically conductive adhesives and grease for thermal management.

The European Customer Service Center in Hilden is the research and development center for customized solutions as well as the main European training center.

LUGATO
Gmbh & Co. KG

Großer Kamp 1
D-22885 Barsbüttel
Phone +49 (0) 40-6 94 07-0
Fax +49 (0) 40-6 94 07-1 10
Email: info@lugato.de
www.lugato.de

Member of IVK, FCIO,
Deutsche Bauchemie e.V.

Company

Year of formation
1919

Size of workforce
About 135

Ownership structure
Gmbh & Co. KG, associated company of the
Ardex Gmbh since 2005

Sales channels
DIY stores

Contact partners
Application technology and sales:
info@lugato.de

Further information
www.lugato.de
www.lugato.com

Range of Products

Types of adhesives
Dispersion adhesives
Cementitious adhesives
MS/SMP adhesives

Types of sealants
Acrylic sealants
Silicone sealants
MS/SMP sealants

For applications in the field of
Household, recreation and office

Mapei Austria Gmbh

Fräuleinmühle 2
A-3134 Nußdorf o.d. Traisen
Phone +43 (0) 27 83-88 91
Fax +43 (0) 27 83-88 91-1 25
Email: office@mapei.at
www.mapei.at

Member of IVK, FCIO

Company

Year of formation
1980

Size of workforce
140 employees

Managing partners
Mapei SpA, Milano, Italien

Subsidiaries
Mapei Kft, Ungarn, Mapei sro,
Tschechische Republik;
Mapefin Austria GmbH

Contact partners
Management:
Mag. Andreas Wolf

Sales management:
Gerhard Tauschmann

Product management:
Stefan Schallerbauer

Range of Products

Types of adhesives
Hot melt adhesives
Reactive adhesives
Solvent-based adhesives
Dispersion adhesives

Types of sealants
Acrylic sealants
Silicone sealants

For applications in the field of
Construction industry, including floors,
walls and ceilings

MAPEI S.p.A.

MAPEI GmbH
IHP Nord – Bürogebäude 1
Babenhäuser Straße 50
D-63762 Großostheim

Via Cafiero 22
20158 Milano, Italy
Phone +39 02/376 731
Fax +39 02/376 732 14
Email: mapei@mapei.it
www.mapei.com

Member of IVK

Company

Year of formation
1937

Size of workforce
more than 10,500

Managing partners
MAPEI S.p.A., Milano, Italy

Ownership structure
family-owned enterprise

Subsidiaries
87 subsidiaries in 35 countries

Sales channels
wholesalers

Contact partners
Management:
Flavio Terruzzi

Range of Products

Types of adhesives
Reactive adhesives
Dispersion adhesives
Pressure-sensitive adhesives

Types of sealants
Acrylics sealants
PUR sealants
Silicone sealants
MS/SMP sealants
Other

For applications in the field of
Construction industry, including floors,
walls and ceilings

merz+benteli ag
more than bonding

Merbenit Gomastit Merbenature

merz+benteli ag
Freiburgstrasse 616
CH-3172 Niederwangen
Phone +41 (31) 980 48 48
Fax +41 (31) 980 48 49
Email: info@merz-benteli.ch
www.merz-benteli.ch

Member of FKS

Company

Organisation
merz+benteli ag was founded in 1918 to supply the Swiss watch industry with innovative adhesives. The enterprise is a family-owned joint-stock company specialised in adhesives and sealants. R&D and production in Switzerland.

Contact
Simon Bienz
Director Marketing & Sales

Distribution partners
DE/AT: Reiss Kraft GmbH
www.reiss-kraft.de / + 49 7253 93 47 65

GR: Dialinas AE
www.dialinas.gr / + 30 210 27 13 333

ES, PT, GB, IE, RU, DK, TR, SE, NO, FI, IT, BE, NL, LU, FR, CZ, SI
Distribution Groupe Europe
www.dge-europe.com / + 31 172 436 361

HU: Güteber Kft.
probond@berenyizoltan.hu / + 36 1 213 5005

SL: Koop KOOP Trgovina d.o.o.
www.koop.si / + 386 (7) 477-8820

IL: Rotal Adhesives & Chemicals Ltd.
www.rotal.com / + 972 9 766 7990

USA, CAN, MEX: Chenso Inc.
www.chenso.com / + 1 336 681 4131

Asia/Pacific: Innosolv Pty.
rick@innosolv.com / + 61 (6) 9846 7871

Range of Products

Range of Products
SMP adhesives and sealants branded as GOMASTIT and MERBENIT.
Bio-based SMP sealants MERBENATURE
Patented MERBENTECH technology

Range of application
Elastic sealing for building construction:
Facades
Interior construction
Floor
Sanitary
Glazing
Roof
Fire protection

Elastic adhesives for industrial constructions:
Industry
Automotive
Marine

 **MINOVA**

Minova Carbotech GmbH
Bamlerstraße 5 d
D-45141 Essen
Phone: +49 (0) 201 80983 500
Email: info.de@minovaglobal.com
www.minovaglobal.com/emea/cis

Member of IVK

Company

Minova International Ltd.
400 Dashwood Lang Road
Addlestone -United Kingdom

Parent-Company
Orica Ltd, Melbourne – Australian

Founding Year
1882 in Dortmund

Employee
1.400

Contact Person
Management:
Michael J. Napoletano, Patrick Langan,
Frank Unterschemmann

Application Technology and Sales:
Herbert Holzer

Further Information
Minova has a 139-year history of developing
and delivering innovative products to the
mining, construction and energy industries.
We are known for our high quality products,
our technical know-how and our problem
solutions. Our innovative, flexible product
portfolio includes a wide range of possible
solutions, that offer you flexibility in your
areas of application at all times.

Range of Products

Types of adhesives
Reactive adhesives

For applications in the field of
Wood/furniture industry,
Construction industry, including floors,
walls and ceilings

We supply an extensive range of products of:

• Anchoring systems of steel and fiberglass
  including accessories for mining and
  tunnelling.
• Injection resins for sealing against gas and
  water ingress, rock and soil consolidation
  and cavity filling.
• Cements and resins for building and sewer
  rehabilitation.
• Adhesives for different floors and coverings.

MÖLLER CHEMIE
CHEMICAL PARTNERSHIP SINCE 1920

Möller Chemie GmbH & Co.KG
Bürgerkamp 1
D-48565 Steinfurt
Phone +49 (0) 2551 9340-0
Fax +49 (0) 2551 9340-60
Email: info@moellerchemie.com
www.moellerchemie.com

Member of IVK

Company

Year of formation
Company was founded 1920

Size of workforce
120

Managing partner
Rainer Berghaus

Ownership structure
Family owned; Private

Subsidiaries
JBC Solution BV

Contact partners
Management:
Rainer Berghaus

Application technology and sales:
Udo Banseberg (TS) und
Frank Dembski (sales)

Further information
Specialized in Additive - Defoamer, Wetting
and Dispersing agent, UV-light Stabilizer,
Silicone surfactants, Rheology agents. Epoxy
curing agent (Reactive diluents, MXDA,
Solvents, Silanes. Hydrophobic and Hydro-
philic fumed Silica of Orisil. Alpha Olefine
of Chevron Phillips (AlphaPlus). Isoparaffine,
Plasticizer, Isotridecanol, etc. Amine and
Metal catalysts, Crosslinker, Other.

Range of Products

Raw materials
Additives
Fillers
Solvents
Polymers
Starch

Equipment, plant and components
Services

For applications in the field of
Paper/packaging
Bookbinding/graphic design
Wood/furniture industry
Construction industry, including floors, walls
and ceilings
Electronics
Mechanical engineering and equipment
construction
Automotive industry, aviation industry
Textile industry
Adhesive tapes, labels
Hygiene
Household, recreation and office

MORCHEM
Your Innovation Partner

MORCHEM, S.A.
Alemania, 18-22
Pol.Ind. Pla de Llerona
E-08520 Les Franqueses del Vallés
Phone +34 93 840 57 02
Fax +34 93 840 57 11
Email: info@morchem.com
www.morchem.com

Member of IVK

Company

Year of formation
1985

Size of workforce
160

Managing partners
CEO: HELMUT SCHAEIDT MURGA

Ownership structure
Family Company

Subsidiaries
MORCHEM GmbH (Germany)/MORCHEM INC (USA)/MORCHEM FZE (UAE)/ MORCHEM SHANGHAI TRADING Co.Ltd. (China)/MORCHEM PRIVATE LIMITED (India)

Contact partners
Management:
Helmut Schaeidt: CEO
Salvador Servera: CCO
Cristina Ventayol: Technical Director
Christoph Moseler: Global Business Director New Markets

Further information
PU based adhesives and coatings for flexible packaging, textile lamination, flat lamination, filtration and other technical laminations. TPUs for printing inks.

Range of Products

Types of adhesives
Hot melt adhesives
Reactive adhesives
Solvent-based adhesives
Dispersion adhesives

Types of sealants
PUR sealants

Raw materials
Polymers

For applications in the field of
Paper/packaging
Electronics
Textile industry

MÜNZING
CREATING ADDITIVE VALUE

MÜNZING CHEMIE GmbH
Münzingstraße 2
D-74232 Abstatt
Phone +49 (0) 71 31-987-0
Fax +49 (0) 71 31-987-202
Email: sales.pca@munzing.com
www.munzing.com

Company

Year of formation
1830

Size of workforce
500

Managing partners
family owned

Subsidiaries
MÜNZING North America,
Bloomfield, NJ, USA
MÜNZING CHEMIE Iberia S.A.U.,
Barcelona, Spain
MÜNZING Poland Sp. z o.o., Poland
MÜNZING International S.a.r.l., Luxembourg
MÜNZING Micro Technologies GmbH,
Elsteraue, Germany
MÜNZING Shanghai Co. Ltd, P.R. China
MAGRABAR LLC, Morton Grove, IL, USA
MÜNZING Mumbai Pvt. Ltd., Mumbai, India
MÜNZING Australia Pty. Ltd., Somersby, NSW,
Australia
MÜNZING Malaysia SDN BHD, Sungai Petani,
Malaysia
MUNZING DO BRASIL, Curitiba/PR, Brazil
Süddeutsche Emulsions-Chemie GmbH,
Mannheim, Germany

Sales channels
direct and via distributors

Contact partners
Application technology:
Peter Bissinger
Tel. No. +49 (0) 71 31-987-174
Email: p.bissinger@munzing.com
Sales:
Dr. Nicholas Büthe
Tel. No. +49 (0) 71 31-987-148

Range of Products

Email: sales.pca@munzing.com
Raw materials
Additives: Defoamers, dispersants, rheology
modifiers, wetting and levelling agents

For applications in the field of
Paper/packaging
Bookbinding/graphic design
Wood/furniture industry
Construction industry, including floors, walls
and ceilings
Electronics
Mechanical engineering and equipment
construction
Automotive industry, aviation industry
Textile industry
Adhesive tapes, labels
Household, recreation and office

www.murexin.com

MUREXIN GmbH
Franz v. Furtenbach Straße 1
A-2700 Wr. Neustadt
Phone +43 (0) 26 22/2 74 01
Email info@murexin.com
www.murexin.com

Member of IVK, FCIO

Company

Year of formation
1931

Size of workforce
400

Ownership structure
Schmid Industrie Holding

Subsidiaries
Croatia, Romania, Czech. Republik,
Hungary, Slovenia,
Slovak. Republic, Poland, France

Contact partners
Management:
Bernhard Mucherl

Range of Products

Types of adhesives
Solvent-based adhesives
Dispersion adhesives
MS adhesives

Types of sealants
Bitumen sealings
Bitumen-free sealings
Joint sealings

For applications in the field of
Construction industry, including floors,
walls and ceilings

Nordmann, Rassmann GmbH
Kajen 2
D-20459 Hamburg
Phone +49 (0) 40 36 87-0
Fax +49 (0) 40 36 87-2 49
Email: info@nordmann.global
Internet: www.nordmann.global

Member of IVK

Company

Year of formation
1912

Size of workforce
460 employees

Managing Board
Dr. Gerd Bergmann, Carsten Güntner, Felix Kruse

Ownership structure
Family-owned

Subsidiaries
Austria, Bulgaria, Czech Republic, France, Germany, Hungary, India, Italy, Japan, Poland, Portugal, Romania, Serbia, Singapore, Slovakia, Slovenia, South Korea, Spain, Sweden, Switzerland, Turkey, United Kingdom, USA

Contact partners
Management: Henning Schild
Phone: +49 (0) 40-3687-248, Fax: +49 (0) 40-3687-72 48
Email: henning.schild@nordmann.global

Sales: Tanja Loitz
Phone +49 (0) 40 3687-313, Fax: +49 (0) 40 3687-7313
Email: tanja.loitz@nordmann.global

Further information
Nordmann has been an independent family business since its foundation in 1912 and belongs to Georg Nordmann Holding AG. The company staffs a total of 460 employees and achieved € 440 million in turnover in 2019. Based in Hamburg, Germany, Nordmann, Rassmann GmbH is the headquarter of Nordmann. For more details please refer to our website at www.nordmann.global.

Nordmann distributes natural and chemical raw materials, additives and specialty chemicals around the world.

We offer partners:
- an innovative product portfolio
- technical know-how and application-related expertise in top industries
- secure and reliable supply-chain solutions
- help with product development
- expertise and support concerning regulatory compliance around the world

Range of Products

Raw materials
Additives: antioxidants/stabilizers, cellulose ethers, dispersion powders, defoamers, PVA, polyethylene oxide, rheology modifiers, polyolefin waxes and copolymers (PE/PP), pigments, impact modifiers, flame retardants and synergists, carbon fibers, functional fillers, starch ethers, surfactants, thickeners

Polymers: styrenic block copolymers (SBS, SEBS, SEP, SIS, SIBS), chloroprene rubber (CR), PVDC, hotmelt polyamide

Resins: hydrocarbon resins, epoxy resins and hardeners, phenolic resins, alkyd resins, tall-oil-rosin-esters, polyterpene resins, terpene phneolics, AMS-resins, AMS phenolics

Reactive components: polyetherpolyols, polyesterpolyols, PTMEG, polycaprolactones, Isocyanates (MDI, TDI), catalysts, reactive diluents, chain extenders, monomers (methacrylates, hydroxymethacrylates, ethermethacrylates, aminomethacrylates)

Plasticizers: mineral oil based process oils, gas-to-liquid (GtL) based process oils, DOTP, DPHP

Dispersions: styrene-acrylate, styrene-butadiene, chloroprene, polyurethane, tall-oil-rosin-ester

For applications in the field of
Adhesive tapes, labels
Automotive industry, aviation industry
Bookbinding/graphic design
Construction industry, including floors, walls and ceilings
Electronics
Household, recreation and office
Hygiene
Mechanical engineering and equipment construction
Paper/packaging
Textile industry
Wood/furniture industry

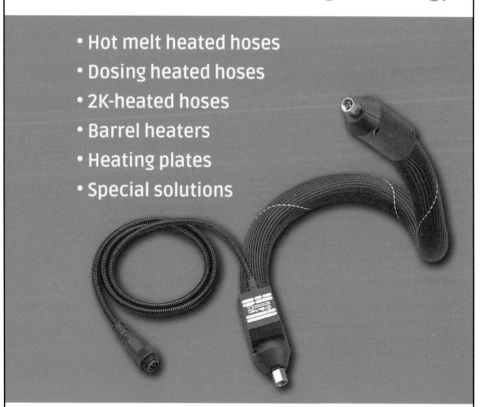

Nynas GmbH
Marktplatz 2
D-40764 Langenfeld
Email: thorsten.wolff@nynas.com
www.nynas.com

Company

Year of formation
1983

Size of workforce
10

Managing partners
Nynas AB Stockholm

Nominal capital
76,693

Subsidiaries
30 offices worldwide

Contact partners
Sales Manager Germany:
Thorsten Wolff

Application technology and sales:
Lars Fischer – process oils
Jan-Peter Laabs – base oils
Nina Lülsdorf – Transformer oils

Range of Products

Process oils for adhesives/hotmelts

For applications in the field of
Paper/packaging
Construction industry, including floors, walls and ceilings
Automotive industry, aviation industry
Textile industry
Adhesive tapes, labels
Hygiene
Household, recreation and office

For stronger adhesives, stick to Nynas

Nynas plasticizers offer superb compatibility with a wide range of polymers and resins. Which means you can produce stronger adhesives with greater cohesion. Our products are available worldwide and designed to comply with the most stringent regulations.

Find what you're looking for at nynas.com

Omya GmbH

Poßmoorweg 2
D-22301 Hamburg
Phone +49 (0) 221-37 75-0
Fax +49 (0) 221-37 75-390
Email: building.de@omya.com
www.omya.de

Member of IVK

Company

Range of Products

Contact
Francois Vielha ber
Email: francois.vielhaber@omya.com

Raw materials
Additives:
Dispersants for water borne systems
Rheology Modifier (PU-/Acrylic thickeners,
Bentonites, Sepiolites)
Super Plasticizer (Dry Polycarboxylate Ether)

Industrial Minerals:
Calcium Carbonates, Dolomite,
Ultrafine PCC
Kaolin, Baryte
Flame Retardants (ATH)
Lightweight Fillers

Polymers:
Vinyl Acetate (VAC)
Acrylic and Styrene-Acrylic Copolymer
Alkyd and Polyester Resins
Styrene Butadiene Rubber (SBR)
Epoxy Resins, Hardeners and Reactive
Diluents
Hot Melt Resins (PA)

For applications in the field of
Paper/packaging
Bookbinding/graphic design
Wood/furniture industry
transportation-/automotive-/aerospace
industry
Construction industry, including floors,
walls and ceilings
Sealants
Casting-, potting materials
Textile industry
Household, recreation and office

Organik Kimya Netherlands B. V.

Chemieweg 7, Havenummer 4206
NL-3197KC Rotterdam Botlek,
Phone +31 10 295 48 20
Fax +31 10 295 48 29
Email: organik@organikkimya.com
www.organikkimya.com

Member of IVK

Company

Year of formation
1924

Size of workforce
500

Managing partners
Simone Kaslowski
Stefano Kaslowski

Nominal capital
100 %

Ownership structure
100 % familiy owned business

Subsidiaries
Distribution worldwide, production sites in the Netherlands and Turkey

Sales channels
Direct sales and through distributors worldwide

Contact partners
Management:
Stefano Kaslowski, General Manager

Application technology and sales:
Oguz Kocak, Sales Manager
Phone: +49 (0) 173-6 52 22 59
Email: o_kocak@organikkimya.com

Range of Products

Types of adhesives
Dispersion adhesives
Pressure-sensitive adhesives

Types of sealants
Acrylic sealants

Raw materials
Polymers

For applications in the field of
Paper/packaging
Bookbinding/graphic design
Wood/furniture industry
Construction industry, including floors, walls and ceilings
Automotive industry, aviation industry
Textile industry
Adhesive tapes, labels

DICHTEN & KLEBEN

Hermann Otto GmbH
Krankenhausstraße 14
D-83413 Fridolfing
Phone +49 (0) 86 84-9 08-0
Fax +49 (0) 86 84-9 08-1840
Email: info@otto-chemie.com
www.otto-chemie.com

Member of IVK

Company

Management board
Johann Hafner
Diethard Bruhn
Matthias Nath
Claudia Heinemann-Nath

Size of workforce
480 employees

Contact
Technical Departement
Phone +49 (0) 8 68 49 08-4300
Email: industry@otto-chemie.de

Sales manager industrial applications:
Marc Wüst
Phone +49 (0) 8 68 49 08-5410

Range of Products

Email: marc.wuest@otto-chemie.de

Types of sealants and adhesives
1-comp. and 2-comp. silicones
1-comp. and 2-comp. polyurethanes
1-comp. and 2-comp. hybrids
acrylates

For applications in the field of
Renewable energies
Domestic appliances industry and
professional cooking technology
Lighting and electronics components
Light construction and composite elements
Laminating an coating
Heating, ventilation and plant construction

Panacol-Elosol GmbH
Member of Hönle Group
Stierstädter Straße 4
D-61449 Steinbach/Taunus
Phone +49 (0) 61 71 62 02-0
Fax +49 (0) 61 71 62 02-5 90
Email: info@panacol.de
www.panacol.com

Member of IVK

Company

Year of formation
1978

Size of workforce
75

Managing Director
Florian Eulenhöfer

Nominal capital
255,645 €

Ownership structure
Panacol-Elosol GmbH is the German subsidiary of Panacol AG in Switzerland. Panacol Group is the subsidiary of Dr. Hönle AG

Sales channels
Sales team for Germany
International distribution network with worldwide sales partners

Contact partners
Sales Director: Eike Leipold

Further information
Panacol is a leading international manufacturer of industrial adhesives as well as medical grade adhesives. In addition to affiliated companies in France, the United States, China and Korea, Panacol provides an international network of distributors, which ensure a personal advisory service around the world. Since January 2008 Panacol has been a member of the Hönle Group and is benefiting from the numerous synergies in the field of industrial UV technology.

Range of Products

Types of adhesives
Reactive adhesives
Anaerobic adhesives
Cyanoacrylates
LED/UVA/visible light curing epoxy and acrylate adhesives
High temperature adhesives
Conductive adhesives
Medical grade adhesives
Structural adhesives

Types of sealants
Epoxies
Acrylics

Equipment
UV and LED equipment for adhesive curing from Hönle UV technology

Market Segments
Electronics/Applications on PCBs
Conformal coatings
Smart card/Die attach
Optics and fibre optics/Active alignment
Optoelectronics
Medical device assembly
Display bonding/Liquid gaskets
Loudspeaker assembly
Appliances
Automotive and Aviation Industry

experience. performance.

Paramelt B.V.
Costerstraat 18
NL-1704 RJ Heerhugowaard
The Netherlands
Phone +31 (0) 72 5 75 06 00
Email: info@paramelt.com
www.paramelt.com

Member of VLK

Company

Year of formation
1898

Size of workforce
522

Ownership structure
privately owned

Subsidiaries
Paramelt Veendam B.V.; Paramelt USA Inc.;
Paramelt Specialty Materials (Suzhou) Co., Ltd.

Sales channels
Sales offices in Sweden, Germany,
Netherlands, UK, France and Portugal and
a network of specialised distributors

Contact partners
Flexible Packaging: Leon Krings
Packaging & Labelling: Kristof Andrzejewski
Construction & Assembly: Wim van Praag

About Paramelt
Paramelt (established in 1898) has grown over
the years to become the leading global specialist in wax based materials and adhesives.
Today Paramelt operates from 8 production
locations around the globe in The Netherlands, USA and China. The company functions
through a series of global business units providing a structured approach to the key market
sectors in which we operate like e.g. packaging
and construction & assembly. For these markets we offer a comprehensive range of waxes,
water based adhesives, hot melt (pressure
sensitive) adhesives, water-based functional
coatings, solvent-based and PU adhesives.
Serviced by both regional sales offices, as well

Range of Products

Types of adhesives
Hot melt adhesives
Reactive adhesives
Solvent-based adhesives
Dispersion adhesives
Vegetable adhesives, casein, dextrin and
starch adhesives
Pressure-sensitive adhesives
Heat seal coatings

Types of sealants
Other

For applications in the field of
Paper/(Flexible) Packaging/Labelling
Construction industry/General Assembly

as a comprehensive network of distribution
partners, our customers can be assured of the
highest levels of local service and support.
Paramelt possess extensive experience in the
design and development of adhesives and
functional coatings to meet critical machine
and application requirements. The company
has built significant knowledge of performance
aspects needed to make our products effective
at all stages of the supply chain. Our products
are backed up by regional laboratories providing
comprehensive application and analytical testing facilities to ensure selection of the most
appropriate adhesive for your application.
Built on a tradition of partnership and trust;
underpinned by detailed knowledge gained
over more than 120 years of operation, Paramelt can offer real benefits to your operation.

PCI Augsburg GmbH

Piccardstraße 11
D-86159 Augsburg
Phone +49 (0) 8 21 59 01-0
Fax +49 (0) 8 21 59 01-3 72
Email: pci-info@PCI-group.eu
www.pci-augsburg.com

Member of IVK, FCIO

Company

Year of formation
1950

Size of workforce
800

Ownership structure
PCI Augsburg GmbH is part of mbcc-group

Subsidiaries
please refer to website for details

Sales channels
indirect / through distributors

Contact partners
Management:
please refer to website

Application technology and sales:
please refer to website

Further information
please refer to website

Range of Products

Types of adhesives
Reactive adhesives
Dispersion adhesives

Types of sealants
Acrylic sealants
PUR sealants
Silicone sealants

Equipment, plant and components
or conveying, mixing, metering and
for adhesive application

For applications in the field of
Construction industry, including floors,
walls and ceilings

PLANATOL®
smart gluing

Planatol GmbH
Fabrikstraße 30 - 32
D-83101 Rohrdorf
Phone +49 (0) 80 31- 7 20-0
Fax +49 (0) 80 31- 7 20-1 80
Email: info@planatol.de
www.planatol.de

Niederlassung Herford:
Hohe Warth 15 - 21
D-32052 Herford
Tel.: +49 (0) 52 21 - 77 01 - 0
Fax: +49 (0) 52 21 - 715 46
Email: info@planatol.de
www.planatol.de

Member of IVK

Company

Year of formation
1932

Size of workforce
120

Ownership structure
Blue Cap AG

Sales channels
Direct sale to industrial customers
Graphical retailers
Representatives abroad worldwide
Subsidiaries abroad

Contact partners
Managing director:
Johann Mühlhauser
Dr. Valentino Di Candido

Range of Products

Types of adhesives
PVAC, EVA, PU & Acryl Dispersions
EVA, PSA, PO Hotmelts
Reactive PUR Hotmelts
Dextrin, Starch and Latex Adhesives
Special-purpose Adhesives

For applications in the field of
Graphics industry
- Fold-gluing
- Bookbinding
- Print finishing

Packaging industry
- Folding box & corrugated cardboard
- End-of-line solutions
- Paper sack production

Wood and furniture industry
- Furniture & kitchens
- Wooden composites
- Design

Special-purpose adhesives for the sandwich construction industry
- Construction industry
- Lightweight construction

Functionalised and custom-made adhesive solutions
- Construction & home
- Automotive
- Textile industry

Wide range of bonding solutions for many industrial applications

Further products
Adhesive application systems for job printing, newspaper printing, gravure printing and digital printing
Adhesive application systems for hot and cold-gluing applications. Copybinder and accessories

POLY · CHEM
POLYMERISATION · SPECIALITY CHEMICALS · SERVICES

POLY-CHEM GmbH
ChemiePark Bitterfeld-Wolfen
Hauptstraße 9
D-06803 Bitterfeld-Wolfen
Phone +49 (0) 3493 75400
Fax +49 (0) 3493 75404
Email: contact@poly-chem.de
www.poly-chem.de

Member of IVK

Company

Year of formation
2000

Size of workforce
50

Sales channels
Direct sales to industrial partners

Contact partners
Management:
Dr. Jörg Dietrich

Application technology:
Dr. Andreas Berndt

Further information
Toll production, Custom Synthesis

Range of Products

We offer to the coating industry, the chemical industry and as well to the segments paints and varnishes.

The present business activities of POLY-CHEM GmbH are
- Solvent based and solvent-free pressure sensitive acrylic adhesives and acrylic polymers in general
- Synthetic rubber pressure sensitive adhesive
- Production of specialty chemicals
- Contract formulations
- Services such as drum filling and tank leasing
- Trading (export/import)

Raw materials
Crosslinker, softening agents, resins, acrylates

For applications in the field of
- Paper/packaging
- Construction industry
- Textile industry
- Adhesive tapes, labels
- Automotive
- Graphic arts

Poly-clip System GmbH & Co. KG
Niedeckerstraße 1
D-65795 Hattersheim am Main
Phone +49 (0) 6190 8886-0
Fax +49 (0) 6190 8886-15360
Email: Vermittlung@polyclip.de
www.polyclip.com/

Member of IVK

Company

Year of formation
1922

Size of workforce
ca. 1,000

Managing partners
Dr. Joachim Meyrahn

Ownership structure
Privately owned

Contact partners
Management:
Björn Arndt

Application technology and sales:
Packaging equipment for flexible cartridges

Further information
Poly-clip System is the largest provider of clip closure systems worldwide and is recognised as world market leader and hidden champion in this sector of the food and packaging industry. The history of the German family company, based in Hattersheim near Frankfurt am Main, stretches back as far as 1922. To date we own more than 800 patents, a fact which underlines our global technology leadership. With an export rate of almost 90 % our clip closure solutions are in use around the world. Everyone is familiar with the ends of a sausage, the 'tails'. Originally developed by us for the meat processing industry and the

Range of Products

Types of adhesives
Reactive adhesives
Solvent-based adhesives
Dispersion adhesives

Types of sealants
Acrylic sealants
PUR sealants
Silicone sealants
MS/SMP sealants

Equipment, plant and components
for conveying, mixing, metering and for adhesive application

For applications in the field of
Construction industry, including floors, walls and ceilings
Mechanical engineering and equipment construction
Automotive industry, aviation industry

---

butchery trade, our clip closure system has long been successfully employed in other food and non-food areas. With our clip-pak technology we offer a sustainable and cost-efficient alternative to traditional cartridges for manufacturers of adhesives and sealants.

Polytec PT GmbH
Polymere Technologien
Ettlinger Straße 30
D-76307 Karlsbad
Phone +49 (0) 72 43-604-40 00
Fax +49 (0) 72 43-604-42 00
Email: info@polytec-pt.de
www.polytec-pt.de

Member of IVK

Company

Year of formation
2005 (1967)

Size of workforce
35

Contact partners
Management:
info@polytec-pt.de

Application technology and sales:
info@polytec-pt.de

Further information
Polytec PT GmbH develops, manufactures
and distributes special adhesives for applications in electronics, electrical engineering
and automotive electronics as well as the
solar industry and the manufacture of smart
cards. In addition to an extensive portfolio
of electrically and thermally conductive
adhesives, transparent and UV-hardening
products, Polytec PT develops tailored
formulations for the most demanding of
adhesive applications.

Range of Products

Range of Products
Electrically conductive adhesives
Thermally conductive adhesives
Adhesives for optical assemblies/fiber
optics
USP Class VI adhesives for medical devices
High temperature adhesives
Potting compounds
UV-curable adhesives
Surface pretreatment devices

For applications in the field of
Electronics
Mechanical engineering and equipment
construction
Automotive industry, aviation Industry

PolyU GmbH
Otto-Roelen-Straße 1
D-46147 Oberhausen
Phone +49 (0) 208 387 64 90
Fax +49 (0) 208 387 64 999
Email: info@polyu.eu
www.polyu.eu

Member of IVK

Company

Year of formation
2017

Size of workforce
15

Nominal capital
500.000 EUR

Ownership structure
100 % subsidiary of PCC SE, Duisburg

Sales channels
directly via PolyU GmbH

Contact partners
Management:
Dr. Klaus Langerbeins, Dr. Johann Klein

Application technology and sales:
Dr. Michael Senzlober (Technical Director),
Ute Dahlenburg (Sales)

Further information
PolyU GmbH is embedded in the PCC SE,
an internationally active group holding
(www.pcc.eu). PolyU has a strong focus on
the development of innovative PU and SMP
systems for the CASE industry.

In close collaboration with customers PolyU
designs and manufactures tailor-made
products.

Range of Products

Types of adhesives
Reactive adhesives

Types of sealants
MS / SMP sealants

Raw materials
Additives:
Inorganic Rheology Modifiers, Silanes,
Flame Retardants

Polymers:
1k PU-Prepolymers, Polyether Polyols,
Polyester Polyols, Polymer Polyols, PUD

For applications in the field of
Wood/furniture industry
Construction industry, including floors, walls
and ceilings

As PolyU is fully integrated within the PCC SE,
we combine and access the potential of our
sister companies in terms of production
facilities, staff and know-how.

To meet the customers' product requirements
and to create novel products from scratch,
PolyU has a highly qualified and motivated
R&D team at their location in Oberhausen
with own laboratory capabilities.

PRHO-CHEM GmbH

Dohlenstraße 8
D-83101 Rohrdorf-Thansau
Phone +49 (0) 80 31-3 54 92-0
Fax +49 (0) 80 31-3 54 92-29
Email: info@prho-chem.de
www.prho-chem.de

Member of IVK

Company

Year of formation
1994

Ownership structure
private property

Contact partners
Management:
Michael Wentz

Application technology:
Otto Kleinhanß

Sales:
Carmen Wolf

Range of Products

Types of adhesives and sealants
Hot melt adhesives
Reactive adhesives
Dispersion adhesives
Pressure-sensitive adhesives
Polyurethane
Silicone
MS/SMP
Epoxy

For applications in the field of
Paper/packaging
Graphic design
Automotive
Adhesive tapes, labels
Industrial applications
Transportation
Aviation

Rain Carbon Germany GmbH

Varziner Straße 49
D-47138 Duisburg
Phone +49 (0) 2 03-42 96-02
Fax +49 (0) 2 03-42 96-7 62
Email: resins@raincarbon.com
www.novares.de

Member of IVK

Company

Year of formation
1849

Sales channels
own sales force
global agent and distribution network

Contact partners
Director Sales/Marketing
Dr. Matthias Heitmann

Sales Manager
Markus Elsner

Range of Products

Raw materials
Hydrocarbon resins:
hydrogenated
pure monomer based
aromatic
aliphatically mod.
phenolically modified
Indene-coumarone

Modifiers:
high boiling solvents

Rampf Polymer Solutions GmbH & Co. KG
Robert-Bosch-Straße 8 – 10
D-72661 Grafenberg
Phone +40 (0) 71 23-9342-0
Fax +49 (0) 71 23-93 42-2444
Email: polymer.solutions@rampf-group.com
www.rampf-group.com

Member of IVK

Company

Year of formation
1980

Size of workforce
100

Managing partners
Dr. Klaus Schamel

Ownership structure
Family owned

Contact partners
Management:
Dr. Klaus Schamel

Application technology and sales:
Felix Neef

Further information
https://www.rampf-group.com/de/
produkte-loesungen/chemical/klebstoffe/

Range of Products

Types of adhesives
1- and 2-component Polyurethane adhesives
2-component Epoxy adhesives
1- and 2-component Silicon adhesives
Thermoplastic and reactive hotmelt adhesives

Types of sealants
PUR sealants
Silicone sealants
MS/SMP sealants

For applications in the field of
Household
Sandwich panels, caravan and truck bodies
Automotive
Filter
Construction
Electronics
Mechanical engineering

Ramsauer GmbH & Co. KG
Sarstein 17
A-4822 Bad Goisern
Phone +43 (0) 61 35-82 05
Fax +43 (0) 61 35-83 23
Email: office@ramsauer.at
www.ramsauer.at

Member of IVK

Company

Year of formation
1875

Head of Business Administration
Klaus Forstinger, Head of Business
Administration

Ownership structure
private

Contact partners
Management:
Andreas Kain, Sales Manager

Further information
Company history:
When Ferdinand Ramsauer purchased a small chalk quarry near Bad Goisern in 1875, he already had all the skills typical of successful people up to our days. He had his mind set on innovation and was fully focussed on achieving his goals. Within less than 20 years, he increased the chalk output of his quarry about 100 times and made "Ischler Bergkreide" ("Mountain chalk from Bad Ischl") a well-known brand name. Ferdinand Ramsauer and his son Josef - whose name our company bears today - were genuine marketing pioneers.
Right from the beginning, mountain chalk was used primarily for the production of putty for glazing. Initially, the raw material was sold to putty manufacturers. In 1950, Ramsauer began manufacturing putty on its own. The evolution from mining operations to sealant manufacturer was complete.
With the introduction of thermally insulating windows, new plastic and elastic sealants were needed. Ramsauer developed the first such modified putties as early as in the fifties. Between 1960 and 1976, sealants were launched under the very well known names of

Range of Products

Types of adhesives
Reactive adhesives
(1 and 2 component solutions)
Solvent-based adhesives
Dispersion adhesives

Types of sealants
Acrylic sealants
Butyl sealants
PUR sealants
Silicone sealants
MS/SMP sealants

For applications in the field of
Wood/furniture industry
Construction industry, including floors, walls and ceilings
Mechanical engineering and equipment construction
Automotive industry, aviation industry
Hygiene and clean room application
Household, recreation and office

E9 M, HB, R68, HV72, and HV76. Simultaneously, Ramsauer started developing the first water-soluble products, known as acrylates. In 1972, Ramsauer started manufacturing sealants on a silicone basis. In 1976, production of PU foam was initiated. A patent for 2-component systems was registered in 1998.
Currently, Ramsauer manufactures top-quality sealants of all systems and modifications, including a new silicone-free sealant on a hybrid basis and a 2-component silicone system.
When looking back at 135 years of company history, there have been many changes, fundamental changes. However what has remained is the broad perspective and focus on innovation that lives on in the present generation in our company.

Renia Gesellschaft mbH
Ostmerheimer Straße 516
D-51109 Köln
Phone +49 (0) 2 21-63 07 99-0
Fax +49 (0) 2 21-63 07 99-50
Email: info@renia.com
www.renia.com

Member of IVK

Company

Year of formation
1930

Workforce
30

Ownership structure
GmbH, family owned

Sales channels
Direct sales to industrial customers,
distributors and agents world-wide

Subsidiaries
SIEMA Vertriebsgesellschaft mbH
Renia USA Inc.

Contact partners
Management:
Dr. Rainer Buchholz

Application technology:
Dr. Julian Grimme

Exportmanager:
Dr. Rainer Buchholz

Range of Products

Types of adhesives
Solvent-based adhesives
Dispersion adhesives

For applications in the field of
Household, recreation and office
Shoe Industry
Health

RUDERER KLEBETECHNIK GMBH

Harthauser Straße 2
D-85604 Zorneding (Munich)
Phone +49 (0) 8106 2421 -0
Fax +49 (0) 8106 2421 -19
Email: info@ruderer.de
www.ruderer.de

Member of IVK

Company

Year of formation
1987

Size of workforce
> 30

Ownership structure
Family company

Sales channels
direct and distribution

Range of Products

Types of adhesives
Hot melt adhesives
Reactive adhesives
Solvent-based adhesives
Dispersion adhesives
Pressure-sensitive adhesives

Types of sealants
Acrylic sealants
PUR sealants
Silicone sealants
MS/SMP sealants

Equipment, plant and components
for surface pretreatment
for adhesive curing
adhesive curing and drying
measuring and testing

For applications in the field of
Wood/furniture industry
Electronics
Mechanical engineering and equipment
construction
Automotive industry, aviation industry
Textile industry
Household, recreation and office

SABA Dinxperlo BV

Industriestraat 3
NL-7091 DC Dinxperlo
Phone +31 315 65 89 99
Fax +31 315 65 32 07
Email: info@saba-adhesives.com
www.saba-adhesives.com

Member of VLK

Company

Year of formation
1933

Size of workforce
170

Managing director
W. de Zwart

Ownership structure
R. J. Baruch, W. F. K. Otten

Subsidiaries
SABA Polska SP. z o. o.
SABA North America LLC
SABA Pacific
SABA China
SABA Bocholt GmbH

Contact:
info@saba-adhesives.com

Further information
Whether you are active in construction or industry, the adhesives and sealants that you use have to meet stringent requirements. SABA is a producer of high-quality and technically progressive adhesives and sealants. With our knowledge of adhesion and sealing, we are happy to help you optimise your processes, so that you produce better end-products or projects at lower total costs and in a safer working environment. In this way, we work together to strengthen your competitive position.

Read more about SABA:
www.saba-adhesives.com

Range of Products

Types of adhesives
Water-based adhesives
Hot melt adhesives
Solvent-based adhesives
Reactive adhesives
Pressure-sensitive adhesives

Types of sealants
Polysulfide sealants
PUR sealants
Silicone sealants
MS/SMP sealants

Equipment, plant and components
for conveying, mixing, metering and
for adhesive application
for surface pretreatment

For applications in the field of
furniture production
mattress production
foam converting
automotive
pvc bondings
building & construction
marine
transport
civil & environmental engineering

Schill+Seilacher "Struktol" GmbH
Moorfleeter Straße 28
D-22113 Hamburg
Phone +49 (0) 40-733-62-0
Fax +49 (0) 40-733-62-297
Email: polydis@struktol.de
www.struktol.de

Member of IVK

Company

Year of formation
1877

Size of workforce
250 employees in Hamburg

Ownership structure
privately owned

Subsidiaries
Schill+Seilacher, Böblingen (Germany)
Schill+Seilacher Chemie GmbH,
Pirna (Germany)
Struktol Company of America, Ohio (USA)
Struktol Co. Company of America, Georgia (USA)

Sales channels
Germany: Direct
International: Distributeurs and Agencies

Contact partners
Dr.-Ing. Hauke Lengsfeld
(General Manager Reactive Polymers & Flame
Retardants)
Phone: +49 (0) 40 733 62 268
Email: hlengsfeld@struktol.de

Sven Wiemer
(Senior Manager Reactive Polymers & Flame
Retardants)
Phone: +49 (0) 40 733 62 125
Email: swiemer@struktol.de

Further information
Manufacturing and development of tailor-made
solutions of exclusive epoxy prepolymers
and 2C epoxy based composite systems in
cooperation with our customers.

Range of Products

Raw materials
Reactive Polymers & Flame Retardants:
Brandnames
Struktol® Polydis®
Struktol® Polycavit®
Struktol® Polyvertec®
Struktol® Polyphlox®

The product ranges Struktol® Polydis® and
Polycavit® are adducts of either rubber or
elastomer modified epoxy resins to improve the
mechanical properties, such as Impact Resistance, T-Peel and Lap Shear next to a general
improvement of the adhesion behavior.

The Struktol® Polyvertec® range includes both
freely combinable and conditional combinations
of resin and hardener that can be used for fiber
composite manufacturing and its processes.
This range also includes a 100 % bio-based poy-
lester resin for sustainable products.

Applications
Epoxy based
(Structural) Adhesives
Castings
Prepregs
Composites (Hand lay-up, RIM/RTM, SMC/
BMC, Pultrusion)
Fiber-Reinforced Plastics
Flame Ratardancy

Schlüter-Systems KG

Schmölestraße 7
D-58640 Iserlohn/Germany
Phone +49 (0) 23 71-971-0
Fax +49 (0) 23 71-971-111
Email: info@schlueter.de
www.schlueter-systems.com

Member of IVK

Company

Year of formation
1966

Size of workforce
more than 1,900 employees worldwide

Managing partners
Schlüter-Systems KG has system alliances with leading construction chemistry and ceramics companies.

Our constructions chemistry partners are: ARDEX GmbH, PCI Augsburg GmbH, Sopro Bauchemie GmbH, MAPEI GmbH, Kiesel Bauchemie GmbH u. Co. KG and SCHÖNOX GmbH, RAK Ceramics GmbH and V&B Fliesen GmbH.

Subsidiaries
Apart from its head office in Iserlohn, Germany, the company has seven subsidiaries in Europe and North America as well as several service bureaus and distribution partners around the world. In total the company employs more than 1,900 people.

Sales channels
B2B

Contact partners
Management:
Günter Broeks (Sales Director)

Application technology:
Björn Kosakowski
(Head of International Technical Network)

Further information
From the Schlüter-SCHIENE to complete systems – with innovative ideas and high-quality products Schlüter-Systems KG is the market leader for many tile related products.

Range of Products

Types of adhesives
Reactive adhesives
Dispersion adhesives

Types of sealants
Other

Raw materials
Polymers

Equipment, plant and components
for surface pretreatment
for adhesive curing
adhesive curing and drying

For applications in the field of
Construction industry, including floors, walls and ceilings
Adhesive tapes, labels
Hygiene

Schlüter-Systems KG with its headquarter in Iserlohn, Germany, offers complete systems with a portfolio of more than 10,000 products and product types and has distribution partners around the world. Among other things, the family-run company offers drainage systems for balconies and terraces, underfloor heating systems, barrier-free showers, construction panels, uncoupling and insulation solutions and a multitude of profiles. Innovative illuminated profile technology and wall heating systems enlarge Schlüter-Systems KG product portfolio.

With more than 1,900 employees in Europe and the USA Schlüter-Systems KG sets standards in creating solutions for tile setters worldwide.

Schomburg GmbH & Co. KG

Aquafinstraße 2 - 8
D-32760 Detmold
Phone +49 (0) 52 31-9 53-00
Fax +49 (0) 52 31-9 53-1 23
Email: info@schomburg.de
www.schomburg.de

Member of IVK

Company

Year of formation
1937

Size of workforce
220 (Germany), 580 worldwide

Managing partners
Albert Schomburg
Ralph Schomburg
Alexander Weber

Nominal capital
3,619 Mio. €

Ownership structure
Family and Management owned

Subsidiaries
31 worlwide:
Poland, Czech Republic, USA, India, Turkey,
Luxemburg, Switzerland, Russia, Netherlands, Slovakia, Italy, etc.

Sales channels
Distribution partners

Contact partners
Management:
Ralph Schomburg
Alexander Weber

Application technology and sales:
Holger Sass
Michael Hölscher

Range of Products

Types of adhesives
Reactive adhesives
Dispersion adhesives
Cement based adhesives

Types of sealants
Acrylic sealants
Polysulfide sealants
PUR sealants
Silicone sealants
Other

Equipment, plant and components
for conveying, mixing, metering and
for adhesive application

For applications in the field of
Construction industry, including floors,
walls and ceilings
Mechanical engineering and equipment
construction

SCIGRIP Europe Ltd.
Unit 22, Bentall Business Park, Glover Road
NE37 3 JD Washington Tyne & Wear, UK
Phone +44 (0) 191 419 6444
Fax +44 (0) 191 419 6445
Email: uk.sales@scigrip.com
www.scigrip.com

Member of IVK

Company

Sales channels
Distribution and direct

Contact partners
Management:
Tim Johnson

Application technology and sales:
Alexander Groeger

Further information
SCIGRIP, a global supplier of smart adhesive solutions, delivers the latest advancements in bonding systems and adhesive chemistries for the most demanding applications.

With a dedicated research and development team, SCIGRIP excel at offering customers unique bonding requirements and help to improve component aesthetics, optimize part design and increase productivity.

SCIGRIP now delivers a broader technology platform than ever before and unique adhesive solutions are available for bonding a range of substrates including metals, thermoplastics and thermoset composites.

SCIGRIP offers powerful, industry-leading technologies in 10:1 and 1:1 methyl methacrylate (MMA), adhesives. All SCIGRIP's formulations are environmentally responsible, low volatile organic compound (VOC) adhesives.

SCIGRIP is committed to manufacturing high quality products and maintains ISO 9001 certification at all its US and European

Range of Products

Types of adhesives
MMA Structural Adhesives

For applications in the field of
Construction industry, including floors, walls and ceilings
Electronics
Mechanical engineering and equipment construction
Automotive industry, aviation industry
Household, recreation and office
Marine
SubSea
Advanced Marine
Renewable Energies
e mobility
Building constructions

---

manufacturing facilities. SCIGRIP has also achieved multiple third party product certifications including Lloyd's Register, DNV-GL GREENGUARD and EN45545-2.

SCIGRIP Adhesive is a wholly owned subsidiary of IPS Corporation, a dedicated, successful adhesives manufacturer for over 70 years. This depth of experience has resulted in an organization that is consistently creating new adhesive chemistries and bonding solutions.

BUILDING TRUST

Sika Automotive Hamburg GmbH
Reichsbahnstraße 99
D-22525 Hamburg
Phone: +49 (0) 40-5 40 02-0
Fax: +49 (0) 40-5 40 02-5 15
Email: info.automotive@de.sika.com
www.sikaautomotive.com

Company

Member of IVK
Year of formation
1928

Size of workforce
260

Subsidiaries
Sister companies in 101 countries

Contact partners
Managing Director:
Heinz Gisel

Export:
Kai Paschkowski

Range of Products

Types of adhesives
Hot melt adhesives
Reactive adhesives
Solvent-based adhesives
Dispersion adhesives
Pressure-sensitive adhesives

Types of sealants
PUR sealants
Other

For applications in the field of
Electronics
Automotive industry
Textile industry
Adhesive tapes & labels

**Creating Solutions
for Increased Productivity**
Sika is supplier and development partner to
the automotive industry. Our state-of-the-art
technologies provide solutions for increased
structural performance, added acoustic
comfort and improved production processes. As a specialty company for chemical
products, we concentrate on our core
competencies: **Bonding – Sealing –
Damping – Reinforcing**
As a globally operating company, we are
partner to our customers worldwide. Sika
is represented with its own subsidiaries in
all automobile-producing countries, thus
guaranteeing a professional and fast local
service.

BUILDING TRUST

Sika Deutschland GmbH
Stuttgarter Straße 139
D-72574 Bad Urach
Phone +49 (0) 71 25-940-761
Fax +49 (0) 71 25-940-763
Email: industry@de.sika.com
www.sika.de/industrie

Member of IVK, FKS, VLK

Company

Year of foundation
1910

Size of workforce
25,000 (worldwide Sika AG),
2,200 Sika Deutschland GmbH

Subsidiaries
in more than 100 countries,
see www.sika.com

Sales channels
direct and distribution

Range of Products

Industry and sealants
PUR adhesives
Reactive adhesives
Dispersion adhesives
Solvent-based adhesives
Hot melt adhesives
Pressure-sensitive adhesives
Epoxy adhesives
Acrylic adhesives and sealants
Laminating adhesives
Silicones
SMP adhesives and sealants

Construction
PUR adhesives and sealants
Polysulfide sealants
Acrylic adhesives and sealants
Silicones
SMP adhesives and sealants
Butyl sealants

For applications in the field of
Mechanical engineering and equipment
construction
Construction industry, including floors,
walls and ceilings
Automotive industry, aviation industry
Wood/furniture industry
Electronics
Adhesive tapes, labels
Household, recreation and office
Photovoltaics
Renewable Energies
Marine
Structural glazing
Facades
Building Constructions
Special and commercial vehicle construction
Automotive repair

BUILDING TRUST

Sika Nederland B.V.
Zonnebaan 56
NL-3542 EG
Phone +31 (0) 30-241 01 20
Fax +31 (0) 30-241 01 20
Email: info@nl.sika.com
www.sika.nl

Member of VLK

Company

Size of workforce
140

Subsidiaries
2

Contact partners
Management:
info@nl.sika.com

Application technology and sales:
info@nl.sika.com

Range of Products

Types of adhesives
1 & 2 Component PU
1 & 2 Component STP
1 & 2 Component Silicone
1 & 2 Component Epoxy
2 Component Acrylics
Hotmelt
Dispersion

Types of sealants
PU
STP
Silicone
Acrylics

For applications in the field of
Automotive
Automotive aftermarket
Marine and offshore
Transportation industry: Trucks, Trailers,
Busses, Special Vehicles
Façade
Fenestration
Solar & Wind Energy
Building Components/Off-site Construction
Sandwich Panels

Sopro Bauchemie GmbH
Biebricher Straße 74
D-65203 Wiesbaden
Phone +49 611-1707-239
Fax +49 611-1707-240
Email: international@sopro.com
www.sopro.com

Member of IVK, FCIO

Company

Year of formation
1985 as Dyckerhoff Sopro GmbH, in the year 2002 change in Sopro Bauchemie GmbH

Size of workforce
334 employees

Managing partners
Michael Hecker, Andreas Wilbrand

Subsidiaries
Germany, Hungary, Switzerland, Netherland, Poland, Austria

Sales channels
through specialised distributors

Contact partners
Management:
International Business
Phone +49 611-1707-239
Fax +49 611-1707-240
Email: international@sopro.com

Further information
Sopro offers a comprehensive range of tile-fixing and building chemicals products. Our clear-cut brand strategy has established us as a leading specialist in this sector. Sopro's wide-ranging product portfolio, featuring a consistently high proportion of new products, addresses the full gamut of tiling and building chemicals applications while quaranteeing a product quality that complies, in all respects, with the strict standards set by professional applicators. The builders' merchants sector represents the only acceptable marketing channel for the Sopro brand, Merchants are able to provide both professional tradesmen and the ambitious private client with the standard of counselling appropriate to our high-grade, technologically advanced products.

Range of Products

Types of adhesives
Cementitious adhesives
Reactive adhesives
Dispersion adhesives

Types of sealants
Acrylic sealants
PUR sealants
Silicone sealants

For applications in the field of
Construction industry, including floors, walls and ceilings

Our product range is divided into thress sectors:
Tile fixing products: Tile Adhesives; Tile grouts; Surface fillers; Primers and bonding agents; Waterproofings; Accessories (impact sound insulation and seperating layer systems); Cleaning, impregnation and maintenance; Tiling tools

Building chemical products: Bitumen products; Screeds, binders and construction resins; Mortar and screed additives; Bedding and multi-purpose mortars; Underground construction, gully-repair and shrinkage compounded grouts; Surface fillers; Concrete repairs; Primers, bonding agents, cement paints and silicia sands

Gardening and landscaping products: Drainage and bedding mortars; Paving grouts; Waterproofings; Surface gradient fillers and rapid set mortars; Cleaning, impregnation and maintenance; Landscaping product systems

Stauf Klebstoffwerk GmbH

Oberhausener Str. 1
D-57234 Wilnsdorf
Phone +49 (0) 27 39-3 01-0
Fax +49 (0) 27 39-3 01-2 00
Email: info@stauf.de
www.stauf.de

Member of IVK, FCIO

Company

Year of formation
1828

Size of workforce
81

Ownership stucture
100 % Family Stauf

Managerial head
Volker Stauf, Wolfgang Stauf,
Dr. Frank Gahlmann

Sales channels

Worldwide distribution, own field service and
distributing warehouses in Germany and other
countries for:
• the wood flooring wholesale
• the floor covering wholesale
• the construction material wholesale
• handcraft enterprises
• contractors

Products
• adhesive systems for floor covering and
  wood flooring
• mounting repair adhesives
• artificial turf adhesives
• primers
• levelling compounds
• underlayments
• surface treatment products
• accessories

Contact partners
Management:
Volker Stauf, Phone +49 (0) 27 39-30 10,
info@stauf.de

Product Technology:
Dr. Frank Gahlmann, Phone +49 (0) 27 39-3
01-1 65, frank.gahlmann@stauf.de

Range of Products

Types of adhesives
Polyurethane adhesives
Silane adhesives
Epoxy adhesives
Solvent-based adhesives
Dispersion adhesives

Types of surface treatment products
waterborne coatings for wood flooring
solvent based coatings for wood flooring
Oils
Care + cleaning systems

For applications in the field of
Construction industry, including floors,
walls and ceilings
• Parquet and wood flooring
• End grain wood blocks
• Textile and elastic floor coverings
• Artificial turf
• Mounting repair
• Industrial application

Further information
www.stauf.de

STAUF is a leading system supplier for flooring
technology. For the safe and durable bonding of wood
flooring and floor coverings we research, develop and
produce innovative adhesive systems on a high-grade
raw material basis. Next to the latest adhesive
systems STAUF offers professionel surface treatment
products for wooden surfaces as well as the whole
bandwidth of products for sub floor preparation and
accessories.

STAUF has remained a family-owned company even
after more than 190 years.

Long time experience as well as continuous advance-
ment in a state of the art production and research
environment ensure the constant top-level product
quality and set the standards for the customers of the
wood flooring and floor covering branch.

STOCKMEIER Urethanes GmbH & Co. KG
Im Hengstfeld 15
D-32657 Lemgo
Phone +49 (0) 52 61-66 0 68-0
Fax +49 (0) 52 61-66 0 68-29
Email: urethanes.ger@stockmeier.com
www.stockmeier-urethanes.com

Member of IVK

Company

Year of formation
1991

Size of workforce
approx. 170

Ownership structure
Member of Stockmeier Group

Subsidiaries
STOCKMEIER Urethanes USA Inc.,
Clarksburg/USA, STOCKMEIER Urethanes
France S.A.S.,Cernay, STOCKMEIER Urethanes
UK Ltd., Sowerby Bridge/UK

Sales channels
direct and distribution

Contact partners
Management:
Markus Lamb

Application technology and sales:
Frank Steegmanns

Further information
Stockmeier Urethanes is a leading international manufacturer of polyurethane systems.
We are a specialized subsidiary of the family-owned Stockmeier Group. With four production plants including R & D in Europe and the USA we are developing and producing polyurethane systems for sports and

Range of Products

Types of adhesives
Reactive adhesives

Types of sealants
PUR sealants

For applications in the field of
Wood/furniture industry
Construction industry, including floors, walls and ceilings
Electronics
Mechanical engineering and equipment construction
Automotive industry, aviation industry

---

leisure flooring, ACE (Adhesives, Coatings & Elastomers) and electrical encapsulation since 1991. In our business unit Adhesives we are producing systems for manufactures in the markets Batteries, Filters, Sandwichpanel, Transportation, Caravan or customized products for other industrial applications.
Our top brands are Stobielast, Stobicast, Stobicoll and Stobicoat. More information: www.stockmeier-urethanes.com

Synthomer Deutschland GmbH
Werrastraße 10
D-45768 Marl
Email: europe.adhesives@synthomer.com
www.synthomer.com

Company

Range of Products

Company

Synthomer is one of the world's leading suppliers of speciality polymers and has leadership positions in many industry segments including adhesives, coatings, construction, technical textiles, paper and synthetic latex gloves. Our polymers help customers to enhance the performance of their existing products and create the next generation products that will fulfil the ever increasingly challenging market needs.

We provide reliable product supply security and customer service worldwide, through a strong network of local commercial and technical service teams, regional R&D and production footprint.

Synthomer has its operational Headquarter in London, UK, and provides customer-focused services from regional centres in Harlow (UK); Marl (Germany); Kuala Lumpur (Malaysia); Atlanta (USA) and Beechwood, OH. It employs more than 4,750 employees across and supply our customers from more than 35 sites.

Contact
Dr. Katja Greiner
Global Technical Service Manager - Adhesives
Phone: +49 (0) 2365-49 9816
Mobile: +49 (0) 171-813 7422
Email: katja.greiner@synthomer.com

Dr. Graeme Roan
Global Marketing Manager - Adhesives
Mobile: +1 864 398 7260
Email: graeme.roan@synthomer.com

Raw materials

Synthomer produces a broad range of dispersions and latexes for tapes, labels, protective films, wood and packaging adhesives, foam adhesives, sealants and more.

Our chemistry range includes pure acrylics, styrene acrylics, vinyl acetates, styrene butadienes, as well as redispersible powders and a range of thickeners and other additives. We also offer silicone free and silicone containing release coatings for select tape applications.

Synthopol Chemie
Alter Postweg 35
D-21614 Buxtehude
Phone +49 (0) 41 61-70 71 962
Fax +49 (0) 41 61-8 01 30
Email: bprueter@synthopol.com
www.synthopol.com

Member of IVK

Company

Year of formation
1957

Size of workforce
190

Ownership structure
Family company

Sales channels
Germany and Europe

Contact partners
Management:
Dr. Rüdiger Spohnholz
Phone +49 (0) 41 61-70 71 160
Dr. Stephan Reck
Phone: +49 (0) 41 61-70 71 130

Sales:
Hubert Starzonek (Commercial Manager)
Phone +49 (0) 41 61-70 71 770

Application technology:
Dipl. Ing. Rainer Jack
Phone +49 (0) 41 61-70 71 171

Further information
Birgit Prüter
Phone +49 (0) 41 61-70 71 962

Range of Products

Raw materials
acrylic emulsions
polyurethane dispersions
saturated polyesters
solvent based and waterbased acrylics

For applications in the field of
construction adhesives
automobile adhesives
textile adhesives
pressure sensitive adhesives

TER Chemicals Distribution Group
Börsenbrücke 2
D-20457 Hamburg
Phone +49 (0) 40-30 05 01-0
Email: info@terhell.com
www.terchemicals.com

Member of IVK

Company

Year of formation
1908

Size of workforce
338

Managing shareholder
Christian Westphal

Sales revenue
350 Mio. €

Subsidiaries
21

Sales channels
Trading, Distribution, Salesforce

Contact partners
Management:
Andreas Früh, CEO

Application technology and sales:
Jens Vinke

Further information
www.terchemicals.com

Range of Products

Types of adhesives
Dispersion adhesives
Hot melt adhesives
Pressure-sensitive adhesives
Reactive adhesives
Solvent-based adhesives
UV-curing adhesives

Types of sealants
Butyl sealants
PIB
PUR sealants

Raw materials
Additives
Fillers
Flame retardants
Nano boron nitride
Nano copper
Nano zinc oxide
Polymers
Resins
Solvents

For applications in the field of
Adhesive tapes, labels
Automotive industry, aviation industry
Bookbinding/graphic design
Construction industry, including floors,
walls and ceilings
Electronics
Hygiene
Paper/packaging
Textile industry
Wood/furniture industry

tesa SE

Hugo-Kirchberg-Straße 1
D-22848 Norderstedt
Phone +49 (0) 40 88899-0
Fax +49 (0) 40 88899-6060
www.tesa.de
www.tesa.com

Member of IVK

Company

Year of formation
tesa AG 2001,
tesa SE since march, 30th 2009

Size of workforce
4,716

Ownership structure
100 % subsidiary of Beiersdorf AG, Hamburg

Subsidiaries
61

Sales channels
Industry (e. g. automotive, building, electronics, print & paper, solar), Food and DIY

Contact
www.tesa.com

Further information
tesa SE is one of the world's leading manufacturers of technical adhesive tapes and self-adhesive system solutions (more than 7,000 products) for industrial and professional customers as well as end consumers. Since 2001, tesa SE (4,716 employees) has been a wholly owned affiliate of Beiersdorf AG (whose products include NIVEA, Eucerin, and la prairie). Applications for various industrial sectors, such as the automotive industry, the electronics sector (e.g. smartphones, tablets), printing and paper, building supply, and security concepts for effective brand and product protection, account for about three-quarters of the tesa Group's sales (2020: 1.325,5 billion euros). tesa also partners with the pharmaceuticals industry

Range of Products

Types of adhesives
Hot melt adhesives tapes
Reactive adhesives tapes
Pressure-sensitive adhesives tapes

For applications in the field of
Paper/packaging
Flexographic printing
Wood/furniture industry
Construction industry, including floors, walls and ceilings
Electronics
Mechanical engineering and equipment construction
Automotive industry, aviation industry
Adhesive tapes, labels
Household, recreation and office

to develop and produce medicated patches. tesa earns just under one-quarter of its sales with products for consumers and professional craftsmen, offering 300 applications for end consumers that amongst others make working in the home and the office easier. tesa became famous for branded products for end consumers - like tesa Powerstrips® or tesafilm®, one of the few brand names to be listed in the Duden dictionary. tesa is active in more than 100 countries worldwide. Find out more about tesa SE at www.tesa.com

Tremco CPG Germany Gmbh

Werner-Haepp-Straße 1
92439 Bodenwöhr
Phone: +49 (0) 9434 208 -0
Fax +49 (0) 9434 208-230
Email: info.de@cpg-europe.com
www.cpg-europe.com

Member of VLK

Company

Year of formation
tremco 1928

Size of workforce
> 1,000

Sales channels
* Distributors (Construction business)
* Direct sales to Key Account Customers in the manufacturing industry and to EIFS and Window & Facade System Providers

Contact partners
Business Unit Industrial Solutions:
Markus Weis

Further information
CPG Europe is one of Europe's leading manufacturers of highperformance sealing and bonding products. Sealing tapes, membranes, adhesives and sealants with particular properties form the core of our extensive range.
We are specialises in finding individual solutions for specific requirements from industrial customers and in optimising their production processes.

www.cpg-europe.com

Range of Products

Types of sealants and adhesives
1-part and 2-part silicones
1-part and 2-part polyurethane
1-part Hybrids
Butyl
Acrylic
Hotmelt

For applications in the field of
Window and Facade Systems
Insulating and Structural Glazing
Exterior Insulation and Finish Systems
Construction, incl. window, facade and interior construction
Automotive Aftermarket and Transportation
Household appliances
Electronics
Ventilation and air conditioning
Photovoltaic

TSRC (Lux.) Corporation S.a.r.l.

39-43 Avenue de la Liberte
L-1931 Luxembourg
Phone +352 262972 60
Mobile: +49 174 3032227
Email: info.europe@tsrc-global.com
www.tsrc.com.tw

Member of IVK

Company

Year of formation
2011
(for the Europe branch in Luxembourg)

Size of workforce
17

Contact partners
Management:
Christian Kafka

Application technology and sales:
Beverley Weaver

Range of Products

Types of adhesives
Hot melt adhesives
Solvent-based adhesives
Pressure-sensitive adhesives

Types of sealants
Other (Styrenic Block Copolymers)

Raw materials
Polymers: Styrenic Block Copolymers
(for applications under 1. & 2.)

For applications in the field of
Paper/packaging
Bookbinding/graphic design
Wood/furniture industry
Construction industry, including floors,
walls and ceilings
Automotive industry, aviation industry
Adhesive tapes, labels
Hygiene
Household, recreation and office

Türmerleim AG

Hauptstrasse 15
CH-4102 Binningen
Phone +41 (0) 61 271 21 66
Fax +41 (0) 61 271 21 74
Email: info@tuermerleim.ch
www.tuermerleim.ch

Member of FKS

Company

Year of formation
1992

Size of workforce
8

Managing partners
Marcel Leder-Maeder

Range of Products

Types of adhesives
Hot melt adhesives
Emulsions
Starch, dextrin and casein adhesives
UF-/MUF-resins

For applications in the field of
Paper/packaging
Labelling
Wood/furniture industry
Tissues

Türmerleim GmbH

Arnulfstraße 43
D-67061 Ludwigshafen
Phone +49 (0) 6 21-5 61 07-0
Fax +49 (0) 6 21-5 61 07-12
Email: info@tuermerleim.de
www.tuermerleim.de

Member of IVK, FKS

Company

Year of formation
1889

Size of workforce
130

Managing directors
Matthias Pfeiffer
Dr. Thomas Pfeiffer
Martin Weiland

Subsidiaries
Türmerleim AG, Basel

Contact partners
Management:
Dr. Jörg Liebe
Josef Karl
Tanguy Trippner
Harald Staub

Application technology and sales:
see Management

Range of Products

Types of adhesives
Hot melt adhesives
Emulsions
Starch, dextrin and casein adhesives
UF-/MUF-resins

For applications in the field of
Paper/packaging
Labelling
Wood/furniture industry
Tissues

 **BOLTON ADHESIVES**

UHU GmbH & Co. KG
Herrmannstraße 7, D-77815 Bühl
Phone +49 (0) 72 23-2 84-0
Fax +49 (0) 72 23-2 84-2 88
E-Mail: info@uhu.de

www.UHU.de
www.UHU-profi.de
www.boltonadhesives.com

Headquarter: Bolton Adhesives
Adriaan Volker Huis – 14th floor
Oostmaaslaan 67
NL-3063 AN Rotterdam

Member of VLK

Company

Year of formation
1905

Size of workforce
Bolton Adhesives > 700 employees

Managing partners
Bolton Adhesives B.V., Bolton Group

Ownership structure
UHU is member of the Bolton Group

Subsidiaries
UHU Austria Ges.m.b.H., Wien (A)
UHU France S.A.R.L., Courbevoie (F)
UHU-BISON Hellas LTD, Pireus (GR)
UHU Ibérica Adesivos, Lda., Lisboa (P)

Contact partners
Managing Director:
Robert Uytdewillegen, Danny Witjes,
Ralf Schniedenharn

Application technology:
Domenico Verrina

Sales:
Stefan Hilbrath

Sales channels
professional trade, hardware stores,
do-it-yourself hypermarkets, modelbuilding
stores, food trade, stationery, department
stores

Range of Products

Types of adhesives
2K-epoxy resin adhesives
Cyanoacrylate adhesives
Hot melt adhesives
Reactive adhesives
Solvent-based adhesives
Dispersion adhesives
Construction adhesives
Sealants

For applications in the field of
Paper/packaging
Wood/furniture industry
Construction industry
Electronics in industry
Automotive industry
Textile industry

UNITECH Deutschland GmbH
Mündelheimer Weg 51a – 53
D-40472 Düsseldorf
Phone +49 (0) 211 51 62 1987
Email: h.balcke@unitech-germany.de
www.unitechcorp.com

Member of IVK

Company

Year of formation
1999

Size of workforce
250

Subsidiaries
South Korea (HQ), Slovakia, Germany,
Turkey, China

Sales channels
Direct / Indirect

Contact partners
Management Director
Hendrik A. Balcke
Email: h.balcke@unitech-germany.de

Head of Product Management
Dr. Benjamin Kraemer
Email: b.kraemer@unitech-germany.de

Range of Products

Types of adhesives
Reactive adhesives
Epoxy adhesives(1C and 2C solution)
Polyurethane(1C and 2C solution)

Types of sealants
PVC-based sealants

Other Products
Composite(EP Prepreg)
Thermal interface material

For applications in the field of
Automotive industry
(Bodyshop, Paintshop and
Interior Shop materials)
Ship building
Electronics
Composites

Uzin Utz

Uzin Utz AG
Dieselstraße 3
D-89079 Ulm
Phone +49 (0) 7 31-40 97-0
Fax +49 (0) 7 31-40 97-1 10
Email: info@uzin-utz.com
www.uzin-utz.com

Member of IVK, FCIO, FKS

Company

Year of formation
1911

Size of workforce
1,335 (2020)

Exectutive Board
Heinz Leibundgut
Julian Utz
Philipp Utz

Ownership structure
Public company

Subsidiaries
Germany, Austria, Belgium, China, Croatia,
Czech Republic, Denmark, France, Hungary,
Indonesia, Netherlands, New Zealand,
Poland, Serbia, Singapore, Slovenia, Sweden,
Switzerland, United Kingdom, USA

Sales channels
Trade, Direct marketing

Contact partners
Head of Sales
Philipp Utz

Head of Research and Development
Dr. Johannis Tsalos

Further information
The Uzin Utz AG is a listed family business
standing for its overall floor expertise. In its
over 100-years history the group of companies has been developing from a small,
regional adhesive manufacturer to one of
the world's leading developers and manu-

Range of Products

Types of adhesives
Reactive adhesives
Dispersion adhesives
Pressure-sensitive adhesives
Cement based adhesives
Adhesives on film carrier

Types of sealants
Dispersion sealants
Reactive sealants
Silicone sealants
Cement based sealants

Equipment, plant and components
For mixing
For adhesive application
For subfloor preparation
For surface finishing

For applications in the field of
Construction industry: installation, renovation and maintainance of all types of floor
coverings.

On walls in the field of tiles and natural stone

facturers of construction chemical system
products, including surface finishing and
machines for tillage. With its outstanding
technology competency Uzin Utz provides
extensive know-how on installations, renovation and maintenance of all types of flooring.
This full-range supplement is marketed worldwide under the brands Uzin, Wollf, Pallmann,
Arturo, codex, RZ and Pajarito.

Uzin Utz
SCHWEIZ

Uzin Utz Schweiz AG
Ennetbürgerstrasse 47
CH-6374 Buochs
Phone + 41 41 624 48 88
Fax + 41 41 624 48 89
Email: ch@uzin-utz.com
www.uzin-utz.com

Member of FKS

Company

Year of formation
1933

Size of workforce
38 (55 with subsidiary)

Managing partners
Vitus Meier

Ownership structure
Public company
A company of the Uzin Utz Group
since 1998

Subsidiaries
DS Derendinger AG, Thörishaus, Switzerland

Sales channels
Direct sales, wholesale distribution, trade

Contact partners
Management:
Vitus Meier, Managing Director

Application technology and sales:
Hans Gallati, Head of Sales and Marketing
Switzerland

Further information
Uzin Utz Schweiz AG stands for concentra-
ted floor competence. Since its foundation
in the year 1933, the company has developed itself to a leading full-range system
supplier for flooring systems in Switzerland.
With the brands UZIN, WOLFF, Pallmann,
codex, Derendinger and collfox, Uzin Utz
Schweiz AG provides a full comprehensive
range.

Range of Products

Types of adhesives
Reactive adhesives
Solvent-based adhesives
Dispersion adhesives
Adhesives on special film carrier

Equiment, plant and components
for conveying, mixing, metering and for
adhesive application
for surface pretreatment
measuring and testing

For applications in the field of
Construction industry, including floors,
walls and ceilings
Transport (railway, ships)

Versalis S.p.A.

Piazza Boldrini, 1
20097 San Donato Milanese, Italy
Email: info@versalis.eni.com
www.versalis.eni.com

Member of IVK through
Versalis International SA -
Zweigniederlassung Deutschland
Düsseldorfer Straße 13
65760 Eschborn, Germany

Company

Year of formation
1957

Size of workforce
5,200

Managing partners
Giovanni Cassuti,
Vice President BU Elastomers

Nominal capital
1.364.790.000 Euro

Ownership structure
Eni S.p.A.

Subsidiaries
see website

Sales channels
Versalis sales network,
see website

Contact partners
Management:
Filippo Forlani,
Sales Manager Elastomers + local sales
network in single countries

Further information
See official website:
www.versalis.eni.com

Range of Products

Types of adhesives
Hot melt adhesives
Solvent-based adhesives
Pressure-sensitive adhesives

Raw materials
Polymers

For applications in the field of
Paper/packaging
Bookbinding/graphic design
Wood/furniture industry
Automotive industry, aviation industry
Adhesive tapes, labels
Hygiene
Household, recreation and office

Vinavil S. p. A.

Via Valtellina, 63
I-20159 Milano
Phone +39-02-69 55 41
Fax +39-02-69 55 48 90
Email: vinavil@vinavil.it
www.vinavil.it

Member of IVK

Company

Year of formation
1994

Size of workforce
> 300

Ownership structure
Mapei S. p. A.

Subsidiaries
Vinavil Americas. Corp.
Vinavil Egypt

Sales channels
> 40 representative commercial offices

Contact partners
Management:
Taako Brouwer
Ing. Silvio Pellerani

Application technology and sales:
Dr. Mario De Filippis
Manfred Halbach
Dr. Fabio Chiozza

Further information
Certified acc. to ISO EN 9001, 14001 and
OHSAS 18001

Range of Products

Types of adhesives
Dispersion adhesives
Pressure-sensitive adhesives

Types of sealants
Acrylic sealants

Raw materials
Polymers:
aqueous polymer dispersions,
solid resins and redispersible powders
RAVEMUL®, VINAVIL®, CRILAT®, RAVIFLEX®
and VINAFLEX® based on vinylacetate, vinylacetate copolymers, vinylacetate ethylene
copolymers, acrylic and styrene acrylic

For applications in the field of
Paper/packaging
Bookbinding/graphic design
Wood/furniture industry
Construction industry, including floors,
walls and ceilings
Automotive industry, aviation industry
Textile industry
Adhesive tapes, labels
Household, recreation and office

VITO IRMEN GmbH & Co. KG
Mittelstraße 74 – 80
D-53424 Remagen
Phone +49 (0) 26 42-4 00 70
Email: info@vito-irmen.de
www.vito-irmen.de

Member of IVK

Company

Year of formation
1907

Size of workforce
85

Managing director
Dr. Michael Büchner

Nominal capital
4,000,000 €

Ownership structure
Limited commercial partnership

Subsidiaries
Representatives in Poland, Austria, Spain,
Netherlands, Czech Republic, Hungary,
Russia

Sales channels
direct and via dealers & distributors

Contact partners
Management:
Dr. Michael Büchner

Application technology and sales:
Marko Rubčić

Further information
www.vito-irmen.de

Range of Products

Self-adhesives tapes coated with
Hot melt adhesives
Solvent-based adhesives
Dispersion adhesives
Pressure-sensitive adhesives

For applications in the field of
- Adhesive tapes, labels
- Automotive industry/aviation industry
- Construction industry, including floors,
  walls and ceilings
- Device for production, storage and trans-
  port of glass
- Electronic Assembly
- Hygiene
- Mechanical engineering and equipment
  construction
- Solar Industry
- Structural Glazing
- Wood/furniture industry

Wacker Chemie AG
Hanns-Seidel-Platz 4
D-81737 Munich
www.wacker.com/contact
www.wacker.com

Member of IVK

Company

Year of foundation
1914

Size of workforce
about 14,300 employees
on December 31, 2020

Ownership structure
Stock corporation ("Aktiengesellschaft")

Subsidiaries
26 production sites

Subsidiaries and sales offices in 32 countries
in Europe, the Americas and Asia.

Range of Products

Raw Materials
Vinyl acetate polymers:
dispersions, dispersible polymer powders
and solid resins (VINNAPAS® and VINNEX®)
Vinyl acetate/ethylene copolymers:
dispersions and dispersible polymer
powders (VINNAPAS® and VINNEX®)
VC copolymers (VINNOL®)
Silicones

Additives:
Pyrogenic silica (HDK®)
Silanes, adhesion promoters and cross-
linkers (GENIOSIL®)
Foam-control agents and silicone
surfactants
Nanoscale silicone particles for modifying
adhesives (GENIOPERL®)

Sealants and Adhesive Grades
RTV-1 Silicones (ELASTOSIL®)
RTV-2 Silicones
LSR Silicones
Silicone gels and silicone foams
UV-curing systems
Hybrid adhesives (GENIOSIL®)

Anspruch verbindet

Wakol GmbH
Bottenbacher Straße 30
D-66954 Pirmasens
Phone +49 (0) 63 31-80 01-0
Email: info@wakol.com
www.wakol.com

Member of IVK, FCIO, FKS

Company

Year of formation
1934

Size of workforce
382

Management
Steffen Acker
Christian Groß (CEO)
Dr. Martin Schäfer

Subsidiaries
Ditzingen/Germany, Austria, Switzerland,
Poland, Italy, USA, China, Brasil

Sales channels
direct distribution, specialised trade

Range of Products

Types of adhesives
Dispersion-based adhesives &
Sealing Compounds
Reactive adhesives
Solvent-based adhesives
Hot melt adhesives
PVC Plastisols
PU Foam

For applications in the field of
Construction industry, including floors,
walls and ceilings
Automotive industry, Aviation industry,
including passenger seats,
Wood/furniture industry, Metal packaging
industry

WEICON GmbH & Co. KG
Koenigsberger Straße 255
D-48157 Muenster
Phone +49 (0) 2 51-93 22-0
Fax +49 (0) 2 51-93 22-2 44
E-mail: info@weicon.de
www.weicon.de

Member of IVK

Company

Year of formation
1947

Size of workforce
306

Subsidiaries
WEICON Middle East L.L.C
WEICON Kimya Sanayi Tic. Ltd. Sti.
WEICON Inc.
WEICON Romania SRL
WEICON SA (Pty) Ltd
WEICON South East Asia Pte Ltd
WEICON Czech Republic s.r.o.
WEICON Iberica S.L.
WEICON Italia S.r.l.

Sales channels
Technical Distributors, Industry directly

Contact partners
Management:
Ralph Weidling
Ann-Katrin Weidling

Application technology and sales:
Holger Lütfring
Technical Project Manager

Vitali Walter
Sales Manager

Range of Products

Types of adhesives
2-comp. adhesives
Basis: Epoxy resins, PUR, MMA
1-comp. adhesives
Basis: Cyanoacrylate, PUR, MMA, POP
Reactive adhesives
Solvent-based adhesives

Types of sealants
PUR-Sealants
Silicone sealants
MS/SMP-Sealants

For applications in the field of
Advertising technique
Automotive industry
Automotive supply industry
Aviation industry
Construction
Construction industry, including floors, walls
and ceilings
Electrical
Electronics
Engine and transmission construction
Household, recreation and office
Marine industry
Mechanical and apparatus engineering
Mechanical engineering and equipment
construction
Metal and plastics industry
On/Offshore
Paper/packaging
Vehicle construction
Wood/furniture industry

**Weiss Chemie + Technik
GmbH & Co. KG**
Hansastraße 2
D-35708 Haiger
Phone +49 (0) 27 73-8 15-0
Fax +49 (0) 27 73-8 15-2 00
Email: ks@weiss-chemie.de
www.weiss-chemie.de

Member of IVK, GEV

Company

Year of formation
1815

Size of workforce
325 members of staff within the group

Managing partners
WBV – Weiss Beteiligungs- und Verwaltungs-
gesellschaft mbH

Subsidiaries
Haiger, Herzebrock, Niederdreisbach,
Monroe NC (USA)

Nominal capital
2 Mio. €

Ownership structure
Family share holders

Management
Jürgen Grimm

Contact partners
Reception: Phone +49 (0) 27 73-8 15-0
Sales: Phone +49 (0) 27 73-8 15-2 02
Technology: Phone +49 (0) 27 73-8 15-2 55
Purchase: Phone +49 (0) 27 73-8 15-2 41

Sales channels worldwide
Own sales force, specialist distributors
industry and trade, customers with private
label

Range of Products

Business division adhesives

Types of adhesives
Reactive adhesives (PUR, Epoxy)
Cyanoacrylate instant glues
Hybrid adhesives (STP/MS)
Solvent-based adhesives
Dispersion adhesives
Pressure sensitive adhesives
Cleaners, solvent- and tenside based

For applications in the field of
Window- and door industry
(plastic materials, metal, wood)
Airtight building envelope according to EnEV
Dry construction
Transportation/commercial vehicles,
Shipbuilding, railway vehicles,
Caravan industry, container construction
Fire protection
Air-conditioning technology and ventilation
engineering
Sandwich- and composite panels
Building trade
Wood-/furniture industry

Business division sandwich elements
Light sandwich constructions as heat- and
sound insulating sandwich elements applied
in fields like doors, windows, gates, booth
constructions, automotive industry etc.

WS INSEBO GmbH

Industriestraße 24
A-2325 Himberg bei Wien
Phone +43 (0) 2235 86 227 - 0
Fax +43 (0) 2235 86 020
Email: office@insebo.com
www.insebo.com

Member of IVK

Company

Year of formation
1973

Size of workforce
60 employees

Managing partners
Alexander Auerbach-Spielberger

Ownership structure
family-owned company

Sales channels
B2B

Contact partners
Management:
ppa. Vesna Rumpold-Bostele

Application technology and sales:
Werner Gassler; Ing. Klaus Maier;
Ivonne Hinzer; Helmut Ruetz;
Michael Schlögl;
Dipl.-Ing. (FH) Thomas Koternetz

Further information
WS INSEBO is who you get in touch with
whenever it comes to building, renovating,
sealing, insulating or gluing. WS INSEBO
offers a broad assortment of PU-foams,
silicones, adhesives, MS-sealants and joint
sealing tapes for every application. The
company located in Himberg is known for
its high-performance and innovative pro-

Range of Products

Types of adhesives
Reactive adhesives
Dispersion adhesives

Types of sealants
Acrylic sealants
Silicone sealants
MS/SMP sealants

Raw materials
Additives
Fillers
Resins
Polymers

For applications in the field of
Wood/furniture industry
Construction industry, including floors, walls
and ceilings
Household, recreation and office

ducts. The products are well approved and
made for everyday use. WS INSEBO places
a lot of importance on the co-operative
partnership with its customers and offers
extensive consultation, overall solutions and
continuous training for its team.

Wöllner GmbH

Wöllnerstraße 26
D-67065 Ludwigshafen
Phone +49 (0) 621 5402-0
Fax +49 (0) 621 5402-411
Email: info@woellner.de
www.woellner.de

Member of IVK

Company

Year of formation
1896

Size of workforce
approx. 150 employees

Subsidiaries
Wöllner Austria GmbH - Fabriksstraße 4-6,
8111 Gratwein-Straßengel, Austria

Sales channels
Direct sales

Contact partners
Management:
Dr. Barbara März

Application technology and sales:
Jörg Batz – Head of Sales Business Unit
CCC; Dr. Joachim Krakehl – Divisional
Head of Technical Marketing Head of Sales
Business Unit ISD

Further information
Wöllner GmbH is one of Europe's leading
suppliers of soluble silicates, process chemicals and special additives for industrial
applications. Being a family enterprise with
over 125 years of experience, we have
in-depth expertise in applied chemistry,
covering the areas of research, development and production.

We develop innovative solutions which are
specially geared towards the chemical,
construction, paint and paper industries as
well as many other sectors.

Range of Products

Types of adhesives
Reactive adhesives
Vegetable adhesives, dextrin and starch
adhesives

Raw materials
Additives

For applications in the field of
Paper/packaging
Wood/furniture industry
Construction industry, including floors, walls
and ceilings

Worlée-Chemie GmbH

Grusonstraße 26
D-22113 Hamburg
Phone +49 (0) 40-7 33 33-0
Fax +49 (0) 40-7 33 33-11 70
Email: service@worlee.de
www.worlee.com

Member of IVK

Company

Year established
1962
(founding as subsidary of E.H.Worlée & Co.
established 1851 as trading company)

Employees
about 300
(productions sites in Lauenburg and Lubeck,
regional sales offices in Germany and affiliates
abroad)

Share holder
Dr. Albrecht von Eben-Worlée
Reinhold von Eben-Worlée

Properties/Estate
owned by the von Eben-Worlée family
Affiliates:
E.H. Worlée & Co. b.V. Kortenhoef (NL)
E. H. Worlée & Co. (UK) Ltd.
Newcastel-under-Lyme (GB)
Worlée Chemie India Private Limited,
Mumbai(India)
Worlée Italia S.R.L. Mailand (I)
Varistor AG, Neuenhof (CH)
Worlée (Shanghai) Trading Co. Ltd.
Shanghai (CN)

Sales Organisation
regional sales organisations in Germany,
affiliates, distributors and agents abroad

Management
Dr. Albrecht von Eben-Worlée
Reinhold von Eben-Worlée
Joachim Freude

Range of Products

Additives
Acrylic resins
Acrylic dispersions
Alkyd resins
Alkyd emulsions
Polyester
Polyester polyols
Maleic resins
Rosin based hard resins, phenol modified
Adhesion promoter
Special Primer
Pigments
XSBR- water based dispersions of carboxylated
styrene-butadiene latex
HS-SBR water based dispersion of
styrene-butadiene copolymer (high solid)
VA – Vinyl acetate dispersions
VAA – Vinyl acetate copolymer dispersions
PVAC – Polyvinyl acetate resins
Styrene acrylic resins
Butadien acrylic resins
Thickeners
Aliphatic isocyanates
Polyisocyanates
Polythiols
Hydro carbon binder resins

WULFF GmbH u. Co. KG

Wersener Straße 3
D-49504 Lotte
Phone +49 (0) 5404-881-0
Fax +49 (0) 5404-881-849
Email: industrie@wulff-gmbh.de
www.wulff-gmbh.de

Member of IVK

Company

Year of formation
1890

Size of workforce
200

Managing partners
Familiy Israel, Mr. Ernst Dieckmann

Sales channels
directly and handling

Contact partners
Management:
Mr. Alexander Israel, Mr. Jan-Steffen Entrup

Application technology and sales:
Mr. Ralf Hummelt, Mrs. Ute Zimmermann,
Dr. Michael Erberich, Mr. Jörg Dronia

Range of Products

Types of adhesives
Dispersion adhesives

Types of sealants
MS/SMP sealants

For applications in the field of
Construction industry, including floors,
walls and ceilings
Textile industry

ZELU CHEMIE GmbH
Robert-Bosch-Straße 8
D-71711 Murr
Phone +49 (0) 71 44-82 57-0
Fax +49 (0) 71 44-82 57-30
Email: info@zelu.de
www.zelu.de

Member of IVK

Company

Year of formation
1889

Size of workforce
50 employees

Contact for Research & Development/ Application Development
Dr Stefan Kissling

Contact for technical sales
Nathalie Uhrich
Mustafa Türken

Sales channels
Direct sales
Distributors
Private labelling

Range of Products

Types of adhesives
Water-based dispersion adhesives
Hotmelt adhesives
Solvent-based adhesives
Pressure-sensitive adhesives
Polyurethane-adhesives
Epoxy-adhesives
Reaction-adhesives

For applications in the field of
Automotive interior
Upholstered furniture
Chairs/office chairs
Foam fabricators
Mattress manufacturing
Automotive and industrial filters
Construction industry
Leather goods manufacturing

Other products
Polyurethane-systems for:
Head and arm rests for automotive interior, bus and train seats, automotive air and industrial filters, seat cusions and arm rests for office chairs, components for noise and vibration reduction, protectors for sport and leisure

Examples of applications
Adhesives:
Bonding of steering wheels
Vehicle headliner
Manufacturing of upholstered furniture and mattresses
Automotive air and industrial filters
Self-bonding products
Heat resistant adhesives for the insulation industry

COMPANY PROFILES

Equipment and
Plant Manufacturing

Baumer hhs GmbH
Adolf-Dembach-Straße 19
47829 Krefeld
Phone +49 2151 4402-0
Fax +49 2151 4402-111
Email: info.de@baumerhhs.com
www.baumerhhs.com

Company

Year of formation
1986

Size of workforce
280 employees worldwide

Managing partners
Baumer Holding AG, Frauenfeld, Schweiz

Ownership structure
Baumer Holding AG, Frauenfeld, Schweiz

Subsidiaries
China, France, India, Italy, Spain, USA,
United Kingdom

Sales channels
Via Headquarter, subsidiaries and dealer
network worldwide

Contact partners
Management:
Percy Dengler, Dr. Oliver Vietze

Application technology and sales:
Director Sales and Service
Roberto Melim de Sousa

Head of Engineering and Technology
Marco Ahler

Further information
Baumer hhs is your leading and competent
partner for industrial solutions in the folding
carton, end-of-line, cigarette packaging,
corrugated, print finishing, braille and
pharmaceutical industries.

Range of Products

System/Processes/Accessory parts/ Services
Gluing systems for cold glue and hot melt
application
Central adhesive feed systems
Quality assurance with sensor and
camera systems
Controllers
High-pressure piston pumps,
diaphragm pumps

For applications in the field of
Paper/packaging
Bookbinding/graphic design
Wood/furniture industry
Corrugated/Folding Carton

Sustainable production with Baumer hhs

Reduce waste and additives, make packaging more sustainable and boost your profitability

rethink. renew. recycle.
hhs
Baumer Group

Let's stick together

baumerhhs.com

bdtronic GmbH
Ahornweg 4
D-97990 Weikersheim
Phone +49 (0) 7934 104-0
Fax +49 (0) 7934 104-371
Email: sales@bdtronic.de
www.bdtronic.com

Company

Year of formation
2002

Size of workforce
460 employees

Managing partners
MAX Automation SE

Subsidiaries
China, USA, Belgium, Italy, UK

Contact partners
Management:
Patrick Vandenrhijn

Application technology and sales:
Andy Jorissen

Further information
With more than 4,000 installed machines, bdtronic is one of the world's leading machine manufacturers for Dispensing, Impregnation, Hot Riveting and Plasma Technology. bdtronic provides entire system solutions for assemply and production automation to all renown manufacturers in the automotive, electric and electronics industry, medical technology and renewable energy. bdtronic employs more than 400 people worldwide at eight subsidiaries in Europe, Asia and the Americas, including five production facilities in Germany, Italy and USA.

Range of Products

Equipment, plant and components
for conveying, mixing, metering and for adhesive application

for surface pretreatment

For applications in the field of
Electronics

Mechanical engineering and equipment construction

Automotive industry, aviation industry

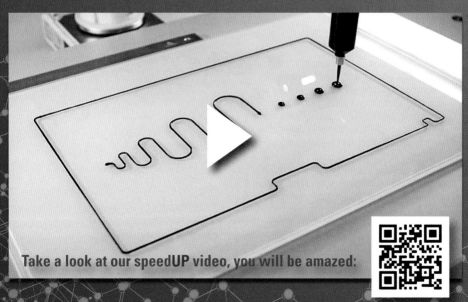

Beinlich Pumpen GmbH
Gewerbestraße 29
D-58285 Gevelsberg
Phone +49 (0) 2332 55 86-0
Fax +49 (0) 2332 55 86-31
Email: info@beinlich-pumps.com
www.beinlich-pumps.com

Company

Year of formation
1951

Managing partners
Jürgen Echterhage

Subsidiaries
SucoVSE France S.A.R.L., UK Flowtechnik Ltd., Oleotec S.r.l (Italy), BEDA Flow Systems Pvt. Ltd. (India), IC Flow Controls, Inc. (USA), E Fluid Technology (Shanghai) Co. Ltd. (China)

Contact partners
Management:
Luigi de Luca, General Manager

Further information
Beinlich Pumpen GmbH is an international supplier of dosing and transfer pumps for industrial applications in process engineering and hydraulic systems. Beinlich offers a large selection of high capacity external and internal gear pumps, high pressure radial piston pumps and progressive cavity pumps and has acquired an extensive technical knowledge in pump technology for more than 70 years. Both the optimal evaluation of individual customer requirements and the precise observation of the markets lead to a continuous development of the products.

On our precise test benches, all pumps are intensively tested for their respective requirements. This way we can assure our customers the highest level of quality and functionality.

Range of Products

Equipment, plant and components
for conveying, mixing, metering and for adhesive application
for surface pretreatment
for adhesive curing

For applications in the field of
Electronics
E-Mobility
Mechanical engineering and equipment construction
Chemical industry
Food industry
Automotive industry, aviation industry
Textile industry
Wood/Furniture industry
Adhesive tapes, labels

ⅠⅠ BÜHNEN

Bühnen GmbH & Co. KG
Hinterm Sielhof 25
D-28277 Bremen
Phone +49 (0) 4 21-51 20-0
Fax +49 (0) 4 21-51 20-2 60
Email: info@buehnen.de
www.buehnen.de

Member of IVK

Company

Year of formation
1922

Size of workforce
102

Ownership structure
Private ownership

Subsidiaries
BÜHNEN Polska Sp. z o. o.
BÜHNEN B. V., NL
BÜHNEN, AT

Contact person
Managing Director & Shareholder:
Bert Gausepohl

Managing Director & Shareholder:
Jan-Hendrik Hunke

Marketing:
Heike Lau

Sales GER, AT, CH, NL, PL, Intl.
Jan-Hendrik Hunke

Range of Products

Distribution channels
Direct Sales, Distributors

Hot Melt Adhesives
The product range includes a
variety of different hot melt adhesives
for almost every application.
Available bases:
EVA, PO, POR, PA, PSA, PUR, Acrylate.
Available shapes:
slugs, sticks, granules, pillows, blocks,
cartridges, barrels, drums, bags.

Application Technology
Hot melt tank applicator systems with
piston pump or gear pump, PUR- and POR-
hot melt tank systems, PUR- and POR-bulk
unloader, hand guns for spray and bead
application, application heads for bead,
slot, spray and dot application and special
application heads for individual customer
requirements, hand-operated glue applicators, PUR- and POR glue applicators, wide
range of application accessories, customer-
oriented application, solutions.

Applications
Automotive, Packaging, Display Manufac-
turing, Electronic Industry, Filter Industry,
Shoe Industry, Foamplastic and Textile
Industry, Case Industry, Construction
Industry, Florists, Wood-Processing and
Furniture Industry, Labelling Industry.

Drei Bond GmbH
Carl-Zeiss-Ring 13
85737 Ismaning, Germany
Phone +49 (0) 89-962427 0
Fax +49 (0) 89-962427 19
Email: info@dreibond.de
www.dreibond.de

Member of IVK

Company

Year of formation
1979

Number of employees
47

Partners
Drei Bond Holding GmbH

Share capital
€ 50,618

Subsidiaries
Drei Bond Polska sp. z o.o. in Kraków

Distribution channels
Directly to the automotive industry
(OEM + tier 1 / tier 2); indirectly
via trading partners as well as select private
label business

Contacts
Management:
Mr. Thomas Brandl

Application engineering, adhesive and
sealants:
Johanna Storm, Dr.Florian Menk

Application engineering, metering technology:
Sebastian Schmidt, Marko Hein

Adhesive and sealant sales:
Thomas Hellstern, Adrian Frey, Stephan Knorz

Metering technology sales:
Sebastian Schmidt, Marko Hein

Additional information
Drei Bond is certified according to ISO 9001-
2015 and ISO 14001-2015

Range of Products

Types of adhesives/sealants
• Cyanoacrylate adhesives
• Anaerobic adhesives and sealants
• UV-light curing adhesives
• 1C/2C epoxy adhesives
• 2C MMA adhesives
• 1 C + 2 C MS hybrid adhesives and sealants
• 1C synthetic adhesives and sealants
• 1C silicone sealants

Complementary products:
• Activators, primers, cleaners

Equipment, systems and components
• Drei Bond Compact metering systems →
  semi-automatic application of adhesives and
  sealants, greases and oils
  Metering technology: pressure/time and
  volumetric
• Drei Bond Inline metering systems →fully
  automated application of adhesives and
  sealants, greases and oils
  Metering technology: pressure/time and
  volumetric
• Drei Bond metering components:
  Container systems: tanks, cartridges, drum
  pumps
  Metering valves: progressive cavity pumps,
  diaphragm valves, pinch valves, spray
  valves, rotor spray

For applications in the following fields
• Automotive industry/automotive suppliers
• Electronics industry
• Elastomer/plastics/metal processing
• Mechanical and apparatus engineering
• Engine and gear manufacturing
• Enclosure manufacturing (metal and plastic)

HARDO-Maschinenbau GmbH
Grüner Sand 78
32107 Bad Salzuflen
Germany
Phone +49 (0) 5222-93015
Fax +49 (0) 5222-93016
Email: coating@hardo.eu
www.hardo.eu

Company

Year of formation
1935

Size of workforce
50 employees

Ownership structure
Family owned

Nominal capital
1.022.500 €

Sales channels
Direct sales and distribution partners worldwide

Contact partners
Managing director:
Ind. & Econ. Engineer Ingo Hausdorf

Sales management:
Hauke Michael Immig

Sales and application technology:
Ralf Drexhage
Reinhard Kölling
Carsten Schoeler

Further information
More than 50 years experience in developing and manufacturing of adhesive application solutions for various industries.

Range of Products

Application systems for
Hotmelts
Pressure sensitive hotmelts
Reactive hotmelts
Dispersion adhesives
Reactive adhesives
Water based primers
Butyl
Bituminous masses
Waxes
Diverse substances

Premelting systems for
Hotmelts
Reactive hotmelts

Roller laminating systems
Tape presses
Plate presses
Winding systems

Customer-specific solutions
HARDO is your ideal partner when it comes to customer-specific solutions. Our team of application technicians and engineers will work together with you to create the optimum machine or system for your work process. Our laboratory will identify the best possible application system for your particular case.

Hilger u. Kern GmbH
Metering and Mixing Technology
Käfertaler Straße 253
D-68167 Mannheim
Phone: + 49 621 3705-500
Fax: + 49 621 3705-200
Email: info@dopag.com
www.dopag.com

Company

Founding Year
1927

Workforce
> 350 worldwide

International Sales and Service
DOPAG Dosiertechnik und Pneumatik AG,
Switzerland
DOPAG S.A.R.L., France
DOPAG UK Ltd., Great Britain
DOPAG Italia S.r.l.
DOPAG (US) Ltd.
DOPAG India Pvt. Ltd.
DOPAG (Shanghai) Metering Technology Co.
Ltd., China
DOPAG Eastern Europe s.r.o., Czech
Republic
DOPAG Korea
DOPAG Mexico Metering Technology SA de CV

Services
- In-house technical centre
- Testing of metering and mixing equipment
- Maintenance
- Repairs
- Spare parts
- Consumables

Further business units
Hilger u. Kern Industrial Technology

Range of Products

Metering and Mixing Technology
- Metering, mixing and dispensing systems for the application of single and two-component materials, e.g. lubricants, adhesives, sealants and potting materials
- Gear pump, piston pump and progressive cavity pump technology
- Automated dispensing systems: standard and line integration-capable production cells, custom-built dosing systems
- Static and static-dynamic mixing systems
- Dynamic mixing head for gasketing, bonding, sealing, and potting applications with PU and silicone

Metering Components and Pumps
- Drum pumps, barrel pumps
- Metering valves, dispensing valves
- Material pressure regulators
- Components for monitoring and control

Applications
- Greasing and oiling
- Bonding and sealing
- Electronic potting
- Composites
- Gasketing
- Thermally conductive materials

Dr. Hönle AG – UV-Technology
Head of Hönle Group
Nicolaus-Otto-Straße 2
D-82205 Gilching
Phone: +49 (0)8105 2083-0
Fax: +49 (0)8105 2083-148
Email: uv@hoenle.com

Company

Year of formation
1976

Annual turnover
93,9 Mio. Euro

Managing Board
Norbert Haimerl
Rainer Pumpe
Heiko Runge

Subsidiaries in the field Adhesives
D: Panacol-Elosol GmbH Deutschland
(Adhesives) - Steinbach/Taunus
uv-technik Speziallampen GmbH
(UV lamps)
F: Honle UV France S.a.r.l., F-Lyon
Eleco Panacol-EFD, F-Gennevillers
Cedex
USA: Panacol-USA, Inc., US-Torrington CT
KOR: Panacol-Korea, KR-Gyeonggi-do
I: Sales Office Hönle Italy
CHN: Hoenle UV Technology (Shanghai)
Trading Ltd., CHN-Shanghai

Sales Contact
Dieter Stirner
Florian Diermeier

Sales channel
Own sales team and sales partners
worldwide

Range of Products

LED-UV/UV technology
for curing UV reactive adhesives, sealing &
potting materials and plastics
for drying and curing UV reactive inks and
coatings
for disinfection of air, water and surfaces
for fluorescence testing
for sun simulation

Adhesives
Development and manufacturing of industrial adhesives via Panacol:
UV and light curing epoxy and acrylate
adhesives
electrical and thermal conductive adhesives
structural adhesives and casting compounds
1K and 2K epoxy resins
special adhesives for medical engineering
and electronics

Application fields bonding technology
Electronics and microelectronics
manufacturing
Conformal coating/chip encapsulation
3-D printing
Precision mechanics
Plants and machinery manufacturing
Automobile and E-mobility
Aerospace industry
Glass industry
Optics
Medical engineering
Photovoltaics

ADHESIVE TECHNOLOGY
as varied as your applications. Since 1965.

H&H Maschinenbau GmbH
Industrieweg 6
D-32457 Porta Westfalica
Phone +49 (0) 571-798 770
Email: info@hh-klebetechnologie.com
www.hh-klebetechnologie.com

Member of IVK

Company

Founded
in 1965

Size of Workforce
60 Team members

Ownership structure
GmbH, family owned

Managing Directors
Michael Hausdorf
Jörg Haubrock

Sale channels
Direct sales and distribution partners
worldwide

Contact partners
Sales management:
Clas-Ole Widderich, Henning Gresmeier
Technical center:
Toni Hausdorf

Company profile
As a well-known, internationally operating,
family-owned company, H&H Adhesive Technology designs, manufactures, and sells high-quality
adhesive application machines and systems
at its Porta Westphalia site in East Westphalia.
More than 50 years of experience in adhesive
application and regard ourselves as your reliable
partner for demanding tasks. Development,
production, installation, service – all from a
single source.

Range of Products

Application System for
Reactive hotmelts
Non-reactive hotmelts
Pressure-sensitive and semi pressure-sensitive
adhesives
Dispersion
Pastes and resins
Glues - cold and hot
Bitumen and butyl
Plastics (PA)
Greases
Waxes and honey

Melting systems for
Granulated hotmelts
Reactive adhesives in a closed-loop system

Solutions for several industries
Automotive industry
Packaging industry
Medical industry
Food industry
Abrasives industry
Glass industry
Electronic industry
Furniture industry
Sports equipment
Building supplies industry
Various industries

Customized systems and solutions
Together with customers, we develop individual
adhesive application system solutions at our
high-performance, in-house technical center.
From project initiation through engineering
to commissioning – We at H&H provide our
expertise to make projects a success, by
customizing our customer's solutions to meet
highly competitive market demands. Delivering
high quality and thus contributing to our customer's success is beyond important to us.

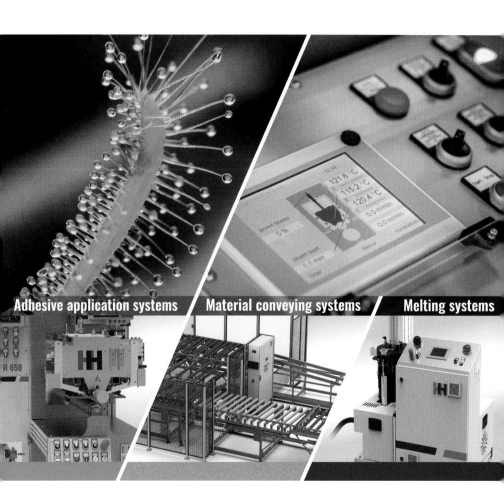

**Innotech Marketing und
Konfektion Rot GmbH**
Schönbornstraße 8 – 8c, D-69242 Rettigheim
Phone +49 (0) 7253-98 88 55-0
Fax +49 (0) 7253-932 40 77
Email: verkauf@innotech-rot.de
www.innotech-rot.de

Member of IVK, GFAV, VTH,
Klebtech Netzwerk, LBZ

Company

Year of formation
1995

Size of workforce
25

Managing partners
Joachim Rapp, Anja Gaber

Nominal capital
100.000 €

Ownership structure
GmbH

Sales channels
technical trade, direct sales, independent sales
representatives

Contact partners
Management: Joachim Rapp, Anja Gaber

Trainings/Consultings: Buruk Sen

Technology/Application equipment: Luca Süß

Application Solutions/Adhesive Accessories:
Alexander Hayes

Further information
Innotech offers a wide range of products and
services and expertise in the field of bonding and
sealing. Innotech has a broad network of more
than 60 adhesive manufacturers in addition to
over 500 dealers worldwide.

Certifications:
• CrefoZert with a credit rating index of 154,
• DIN EN ISO 9001: 2015

The company releases the Manual Adhesive
Application Almanac annually. It serves as a database and collection of knowledge on all subjects
related to adhesive bonding. The aim of the almanac is to provide fundamental information about
various dispensers, cartridges, static mixers,
nozzles and all other adhesive accessories while
further elaborating upon practical adhesive
information and gluing tips. Apart from this,

Range of Products

Applicators
As distributor and service partner of international
leading sealant and adhesive manufacturers,
Innotech offers more than 750 models of applicators and dispensing tools – quality made in
Europe and USA- with more than 200.000 spare
parts. This involves the repair and warranty
service as well as the distribution of specialized
tools, just-in-time delivery, consumer parts and
accessories.

Adhesive accessories
Wide assortment and worldwide dispatch of mixers and nozzles and surface treatment from all
leading manufacturers.

___

numerous industry experts are represented with
their specialist contributions being fully detailed.

Services
Support and advice for applicators, Repair service
for applicators, Support, sample shipping for
adhesive manufacturers, solutions and training,
Cooperation partner with IFAM – Training Center
for European Adhesive Specialist (EAS) and
European Adhesive Bonder (EAB), Conduction of
training sessions of European Adhesion Specialist
(EAS) & European Adhesive Bonder (EAB),
Logistic and transportation, Sample logistics,
Development of special / tailormade nozzles,
Product innovations

Trade
Special applicators, Cartridges
Mixers and nozzles, Adhesive supplies, Cleaners,
Primer equipment, Surface treatment uvc air
disinfection

Production
Marketing bonding, Refilling service
(in different quantities), Test pieces, Packaging/
Assembling

IST METZ GmbH
Lauterstraße 14 – 18
D-72622 Nürtingen
Phone +49 (0) 0 70 22 - 6 00 20
Fax +49 (0) 0 70 22 - 6 00 276
Email: info@ist-uv.com
www.ist-uv.com

Member of IVK

Company

Year of formation
1977

Size of workforce
550 employees worldwide

Ownership structure
GmbH

Subsidiaries
eta plus electronic GmbH
Integration Technology Ltd.
IST France sarl
IST Italia S.r.l.
IST (UK) Limited
IST Nordic AB
IST Benelux B.V.
UV-IST Ibérica SLU
IST America - U.S. Operations, Inc.
IST METZ SEA Co., Ltd.
IST METZ UV Equipment China Ltd. Co.
IST East Asia Co., Ltd.
S1 Optics GmbH

Contact partners
Management:
Christian-Marius Metz

Application technology and sales:
Arnd Riekenbrauck

Range of Products

Equipment, plant and components
for adhesive curing
adhesive curing and drying
measuring and testing

For applications in the field of
Paper/packaging
Building industry
Construction industry, including floors,
walls and ceilings
Electronics
Automotive industry, aviation industry
Adhesive tapes, labels

Nordson Deutschland GmbH
Heinrich-Hertz-Straße 42
D-40699 Erkrath
Phone +49 (0) 211 92 05-0
Fax +49 (0) 211 25 46 58
Email: info@de.nordson.com
www.nordson.com

Company

Year of formation
1967

Size of the workforce
450 employees

Shareholders
Nordson Corporation, USA

Contact partners
Board of management:
Michael Lazin, Ben Peuten,
Gregory Paul Merk

Sales
General sales manager: Olaf Hoffmann
OEM support: Olaf Hoffmann
Packaging/assembly applications:
Olaf Hoffmann
Industrial applications: Jörg Klein
Nonwoven: Kai Kröger
Industrial Coating Systems: Ralf Scheuffgen
Automotive/Sealant Equipment: Volker Jagielki

Sales routes
Through field service staff of Nordson
Deutschland GmbH

Further information
Development centres and production facilities
(ISO-certified) in the USA and Europe, over 7,500
employees, subsidiaries on every continent. In
cooperation with the customer, Nordson develops
complete solutions with integrated systems
and matching components which grow with the
customers' demands.

Range of Products

Equipment and systems for the application of
adhesives and sealants and for surface finishing
with paints, lacquers, other liquid materials or
powders. Nordson systems can be integrated into
existing production lines.

Packaging and assembly applications
Complete adhesive application systems (hot
melt/cold adhesive) for integration into packaging lines. In the field of assembly applications,
Nordson optimises production processes in many
different branches of industry.

Industrial applications
Adhesive and sealant applications for a wide
variety of branches of industry, e.g. filter bonding,
sealing of vehicle rear lights and mattress bonding. Wood processing, e.g. profile wrapping, edge
banding and post-forming.

Nonwoven
Nonwoven (tailored systems for the application
of adhesive and super-absorbent powder for the
production of babies' nappies, slip inlays, sanitary
towels and incontinence articles).

Powder & Liquid Coating
Applications and systems for coating with paints,
lacquers, other liquid materials and powders.

Electronics
Automatic coating and metering plants for the
electronics industry for precise application of
adhesives, grouting compounds, solder pastes,
fluxes, protective lacquers, etc.

Automotive
Engineered systems for the application of structural adhesives and sealants in harsh automotive
manufacturing environments.

Battery Manufacturing
Dispensing systems for 1-part and 2-part
materials used in storage or vehicle battery cell
manufacturing.

Plasmatreat GmbH
Queller Straße 76 – 80
D-33803 Steinhagen
Phone +49 (0)5204-9960-0
Fax +49 (0)5204-9960-33
Email: mail@plasmatreat.de
www.plasmatreat.com
www.book-a-demo.com

Member of IVK

Company

Founded in
1995

CEO
Christian Buske

Number of employees
Approx. 240 (worldwide)

Contact
Sales Manager Germany
Joachim Schüßler

Head of Application Management
Lukas Buske

Subsidiaries & Partners
17 subsidiaries in 11 countries and
numerous partners worldwide

Technology Centers
Germany, USA, Canada, China, Japan

Further information
Atmospheric pressure plasma is a key technology for the environmentally friendly and highly
effective pretreatment and functional coating of
material surfaces. Plasmatreat became the
first company in the world to integrate plasma
into series production processes 'inline' under
normal pressure, thus making it suitable for use
on an industrial scale. The patented Openair-
Plasma® technology is now used worldwide in
virtually all sectors of industry. Plasmatreat
invests around twelve percent of its annual turnover in research and development. The company collaborates with the German Federal

Range of Products

**Systems/processes/accessories/
services**
Surface pretreatment: Microfine cleaning
activation and functional nano coatings

Areas of application
Packaging
Furniture
Glass processing, windows
Electronics
Machinery and appliance manufacturing
Automotive engineering
Truck construction
Aerospace industry
Shipbuilding
Medical engineering
New energies (solar technology, wind
power, electromobility)
Consumer goods
Textiles
Adhesive tapes, labels
Hygiene applications
Household appliances, white goods

Ministry of Education and Research (BMBF) on
numerous research projects and also works
closely with the Fraunhofer Institutes and other
leading research institutes and universities
around the world.

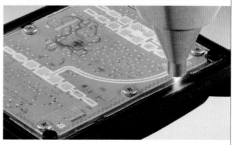

Reinhardt-Technik GmbH
a Member of WAGNER GROUP
Waldheimstraße 3
D-58566 Kierspe
Phone +49 (0) 2359 666-0
Fax +49 (0) 2359 666-129
Email: info-rt@wagner-group.com
www.reinhardt-technik.com

Company

Year of formation
1962

Size of workforce
About 80 employees

Sales Director
Christian Hose (Commercial Director)
Axel Huwald (Manager Sales EMEA)

Direct Sales
Sales representatives, agents and distributors

Company Profile
Reinhardt-Technik GmbH specializes in the areas of
bonding, sealing and casting including injection molding.
The company offers a comprehensive range of machines
for metering and mixing technology, which process cold
or heated 1K materials and multi-component liquid
plastics such as polyurethanes, polysulphides, epoxies,
silicones and LSR (Liquid Silicone Rubber). All common
process technologies are offered - from pneumatically or
hydraulically driven piston pumps, gear metering systems
to electrically controlled shot metering systems. In
addition, Reinhardt-Technik is a solution provider and
supplies complete custom-engineered systems.

Range of Products

Metering and Casting Systems
Highly precise and reliable dosing systems for a variety
of applications:
• Shot metering systems
• Gear metering systems

Automated Production Cells together with partners
Convenient, simple as well as safe operation of the
dosing and mixing systems along with integration with
upstream and downstream processes:
• Standardized robot cells
• Individual and application-specific manufacturing cells

Material Preparation and Feeding
Modular as well as configurable processing and conveying
systems:
• Agitators for fast-sedimenting materials
• 1K and 2K systems
• Optional booster addition
• 20 and 200 litre feeding units with robust working
  cylinder in combination with powerful chop check
  pumps for low viscosity materials
• Optional pressure containers
• Reliable material discharge with low residual material

Mixing Systems
Low-maintenance and powerful mixing systems for
individual processing:
• Static mixing systems
• Dynamic mixing systems
• Incl. snuff back valve

Customer Service
• Technical application centre
• Training
• Hotline
• Remote maintenance
• Online shop (spare and wearing parts)
• High spare part availability
• Service worldwide
• Equipment modernization and modification
• Process optimization

Reka Klebetechnik GmbH & Co. KG
Siemensstraße 6
D-76344 Eggenstein-Leopoldshafen
Phone +49 (0) 721 97078-30
Fax +49 (0) 721 705069
Email: adhaesion@reka-klebetechnik.de
www.reka-klebetechnik.de

Member of IVK

Company

Year of formation
1977

Sales channels
Direct sales and sales via distributors and partners

Contact partners
Management:
Herbert Armbruster

Application technology and sales:
Katharina Armbruster

Further information
Since 1977 Reka Klebetechnik offers its customers in over 70 countries reliable quality from Germany. In Eggenstein near Karlsruhe, Germany, Reka's hand-operated hotmelt adhesive applicators are developed, produced and sold.

The devices are suitable for processing hotmelt adhesives in almost all forms available on the market. They are just as appreciated in industry as they are by craftsmen and packaging service providers. Whether in production, assembly or sealing – most modern industrial products require adhesives.

The steady growth of the worldwide distributor and service network and the constant improvement of the products strengthen the owner-managed family business. Reka Klebetechnik focuses on long-term business relationships and durable products that are easy to maintain.

With its competent and motivated team, decades of market experience and

Range of Products

Types of adhesives
Hot melt adhesives
Pressure-sensitive adhesives

Types of sealants
Other

Application Systems
Industrial Bulk Glue Guns using compressed air;

Industrial Bulk Glue Guns without compressed air

Pneumatic Industrial Glue Guns for 310 ml aluminum cartridges

(single-component adhesives)

For applications in the field of
Paper/packaging
Bookbinding/graphic design
Wood/furniture industry
Construction industry, including floors, walls and ceilings
Electronics
Mechanical engineering and equipment construction
Textile industry
Household, recreation and office

cooperation with renowned adhesive manufacturers, Reka offers its customers solutions tailored to the customers needs. Reka helps you to select the suitable products for your application.

Robatech AG
Pilatusring 10
CH-5630 Muri AG
Phone (+41) 56 675 77-00
Fax (+41) 56 675 77-01
Email: info@robatech.ch
www.robatech.ch

Member of IVK

Company

Year of formation
1975

Size of workforce
More than 650 employees all over the world

Shareholder
Robatech AG, CH-5630 Muri, Switzerland

Ownership structure
Robatech AG, CH-5630 Muri, Switzerland

Subsidiaries and agencies
Represented in more than 80 countries
worldwide
Germany: Robatech GmbH,
Im Gründchen 2, D-65520 Bad Camberg
Phone +49 (0) 64 34-94 11-0
Fax +49 (0) 64 34-94 11-22

Channel of distribuition
Via Head office, subsidiaries and agencies

Contact partners
Management:
Robatech AG, Switzerland:
Martin Meier
Robatech GmbH, Germany:
Eberhard Schlicht, Andreas Schmidt

Application technology and sales:
Robatech AG, Switzerland:
Harald Folk, Sales Director
Kishor Butani, Sales Director
Kevin Ahlers, Marketing Director
Robatech GmbH, Germany:
Eberhard Schlicht, Managing Director

Further information
Production facility in Germany, Hong Kong
and Switzerland

Range of Products

Product and sales program of the company
- Glue application system with piston pumps and gear pumps for hotmelt and dispersions, inclusive necessary equipment.
- Small hotmelt application systems up to 5 liter tank capacity
- Medium hotmelt application systems from 5 to 30 liter tank capacity
- Big hotmelt application systems from 55 to 160 liter tank capacity
- Hotmelt application systems for PUR-hotmelts from 3 to 30 liter tank capacity
- Drumunloaders from 50 to 200 liter tank capacity
- Application technics: Bead application, Surface coating, Spray application, spiral application
- Electrical timing and electrical metering
- Cold glue application systems: Pressure tanks and pump systems

Robatech offers a solution for several industries
Packaging Industry
Converting Industry
Graphic Industry
Woodworking Industry
Building Supplies Industry
Automotive Industry
Various Industries

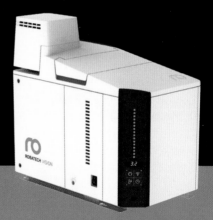

Rocholl GmbH
Industriestraße 28
D-74927 Eschelbronn
Phone +49 (0) 6226 93330 0
Fax +49 (0) 6226 93330 10
Email: post@rocholl.eu
www.rocholl.eu

Member of IVK

Company

Year of formation
1977

Ownership structure
Owner/managing director

Sales channels
direct

Contact partners
Management:
Dr. Matthias Rocholl

Application technology and sales:
Dr. Matthias Rocholl

Range of Products

Equipment, plant and components
measuring and testing

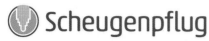

Scheugenpflug
Part of the Atlas Copco Group

Scheugenpflug GmbH
Gewerbepark 23
D-93333 Neustadt
Phone +49 (0) 94 45-95 64-0
Fax +49 (0) 94 45-95 64-40
Email:
sales.de@scheugenpflug-dispensing.com
www.scheugenpflug-dispensing.com

Company

Year of formation
1990

Size of workforce
More than 600 (worldwide)

General Manager
Olaf Leonhardt

Contact
Email: sales.de@scheugenpflug-dispensing.com
www.scheugenpflug-dispensing.com

Subsidiaries
Atlas Copco Industrial Technique (Shanghai) Co.,
Ltd., China
Email: info@scheugenpflug.com.cn
Scheugenpflug Inc., USA
Email: sales.usa@scheugenpflug-dispensing.com
Scheugenpflug México, S. de R.L. de C.V., Mexico
Email: sales.mx@scheugenpflug-dispensing.com
Scheugenpflug S.R.L., Romania
Email: info.ro@scheugenpflug-dispensing.com

Sales Partners Worldwide
See Webpage > Company > Locations & Sales
Partners

Company Profile
Technology that sets standards: With more than
30 years of experience, Scheugenpflug is one of the
leading manufacturers of innovative dispensing technology. With an additional core competency in process
automation, the product and technology range extends
from cutting-edge material preparation and feeding
units and high performance dispensing systems to
custom-tailored, modular production lines for a wide
range of applications. Scheugenpflug systems are
used for the production of electronic components,
especially in the automotive, industrial, medical and
consumer segments. The company has additional
locations in China, the USA, Mexico and Romania as
well as numerous service locations and sales partners
all over the world.

Range of Products

Dispensers
• Volumetric piston dispensers, with application
specific design
  – E.g. for thermally conductive adhesives or as
alternating version
• Gear pump dispensers

Material Preparation and Feeding
• Feeding from cartridges
• Feeding from hobbocks
• Systems for self-leveling potting media

Systems for Atmospheric Dispensing
• Manual work stations with stand or for
integration
• ProcessModules for integration
• Dispensing cells

Vacuum Potting Systems
• Vacuum chambers

Customized Systems and Solutions
Customized automation solutions for various adhesive bonding, sealing and potting tasks, based
on the Scheugenpflug modular system (scalable
design), heat dissipation

Services
• Technology Center/Dispensing tests
• Academy
• Technical Services
• After Sales Services/Spare parts
• Rental equipment

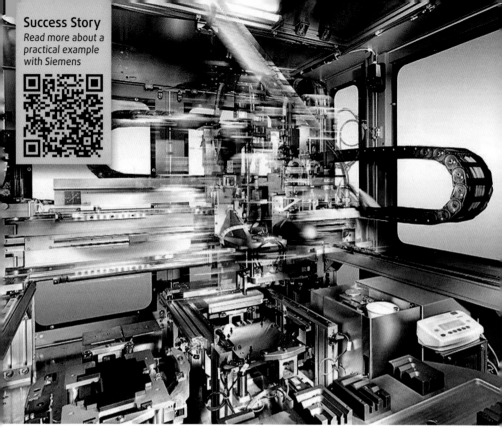

Sulzer Mixpac Ltd.
Ruetistraße 7
CH-9469 Haag
Phone +41 (0) 81 414 70 00
Email: mixpac@sulzer.com
www.sulzer.com

Company

Year of formation
Sulzer Mixpac is part of Sulzer's Applicator Systems (APS) division, which is planned to be spun off from Sulzer Group and renamed medmix after its listing on the SIX Swiss Exchange. The spin-off is planned for H2/2021 and subject to the approval of Sulzer's shareholders at the extraordinary general meeting planned for Q3/2021.

Size of workforce
1,000 employees worldwide

Distribution
Direct distribution to producers of adhesives. toll filler and distribution partners for official trade

Contact partners
Switzerland:
Sulzer Mixpac Ltd.
Email: mixpac@sulzer.com

China:
Sulzer Mixpac China
Email: mixpac@sulzer.com

Untited Kingdom:
Sulzer Mixpac UK
Email: mixpac@sulzer.com

United States:
Sulzer Mixpac USA Inc.
Email: mixpac@sulzer.com

Range of Products

Manufacturer and supplier of metering, mixing, coatings and dispensing systems for reactive multi component material, offering comprehensive systems for various cartridge based 2-K applications.

Cartridge based 2-K systems with different volumes between 2.5 ml and 1.500 ml, with mixing rations of 1:1, 2:1, 3:1, 4:1 and 10:1
Manual, pneumatic and electric dispensers for 1K and 2K adhesives and sealant applications.
Full range of mixers for cartridge and dispensing machine applications
Brands: MIXPAC™, MK™, COX™

SOLDERING & DISPENSING

unitechnologies
THE ART OF PRECISION

Unitechnologies SA - mta®
Bernstrasse 5
CH-3238 Gals
Phone +41 32 338 80 80
Email: info@unitechnologies.com
www.unitechnologies.com
https://mtaautomation.com/dosieren/

Company

Year of formation
1966

Size of workforce
40

Ownership structure
AG (Aktiengesellschaft)

Subsidiaries
Deutschland – mta automation gmbh,
Bert-Brecht-Weg 6,
71549 Auenwald-Unterbrüden,
USA – mta automation inc., 123 Popular
Pointe Drive, Suite E, Mooresville, NC 28117

Contact partners
Management: Alessandro Sibilia (CEO)
Application technology and sales:
Stefan Eidam (Senior Sales Manager)

Company profile
Unitechnologies is a process specialist in the volumetric dispensing and selective soldering domains, with over 50 years of experience in precision automation.

mta® high precision volumetric dispensing machines are used for liquid and high-viscosity materials with mono- or two-components in a variety of dispensing processes. As a result, the product range includes simple manual systems, semi-automated table robots as well as fully automated production cells, achieving a perfect repeatability of micro-volumes, starting at 0.1 mm³.

Range of Products

Volumetric dispensing heads
- Piston-cylinder pump dispensers
- Eccentric screw pump dispensers
- Jet dispensers
- Cartridge dispensers

Processes and application fields
Gluing, adhesive bonding, sealing, potting, conformal coating, greasing, dam & fill, glob top and underfill in industries such as electronics, optoelectronics, automotive, watchmaking, medical and pharma.

Personalized service
Unitechnologies focuses on high-quality applications and relies on the expertise of highly-qualified specialists to conduct dispensing process validation in its laboratory, which is equipped with state-of-the-art equipment. This includes standard automated cells, table-top robots and all above mentioned dispensing heads. Other equipment is also available such as certified precision balances, ovens, UVcuring devices and many more.

The aim of Unitechnologies test program is essentially to check the feasibility of the required process thus enabling to provide a process guarantee for each application. The customer benefits by receiving a proof of process prior to investing in the production equipment.

ViscoTec Pumpen- u. Dosiertechnik GmbH
Amperstraße 13
84513 Töging a. Inn
Phone +49 (0) 8631 9274-0
Email: mail@viscotec.de
www.viscotec.de

Company

Year of formation
1997

Size of workforce
260 employees (worldwide)

Managing partners
Georg Senftl
Martin Stadler
Franz Kamhuber

Subsidiaries
- Töging am Inn, Germany (Headquarter)
- Georgia, USA
- Singapore
- Shanghai, China
- Pune, India
- Mérignac, France

Sales channels
worldwide

Further information
ViscoTec manufactures systems required for conveying, dosing, applying, filling, and emptying medium to high-viscosity materials. The fluids can be conveyed and dosed almost pulsation-free and with extremely low shear.

Range of Products

Equipment, plant and components
Systems for conveying, dosing, applying, filling and emptying fluids ranging from medium to high viscosity:
- Dispenser and dosing systems (applying and administering, 1- and 2-component dispensing, potting, filling, process dosing, spraying)
- Emptying and supplying systems
- Treatment systems
- Complete systems
- Dedicated equipment

For applications in the field of
- Automotive
- Aerospace
- Electronics
- General industry
- Dental
- Food
- Cosmetics
- Pharmaceuticals
- Medical technology
- Consumer Goods
- 3D printing
- E-mobility

VSE Volumentechnik GmbH
Hönnestraße 49
D-58809 Neuenrade
Phone +49 (0) 2394 616-30
Fax +49 (0) 2394 616-33
Email: info@vse-flow.com
www.vse-flow.com

Company

Year of formation
1989

Managing partners
Jürgen Echterhage, Axel Vedder

Subsidiaries
SucoVSE France S.A.R.L., UK Flowtechnik Ltd., Oleotec Srl. (Italy), BEDA Flow Systems Pvt. Ltd. (India), IC Flow Controls, Inc.(USA), E Fluid Technology (Shanghai) Co. Ltd. (China)

Further information
VSE Volumentechnik stands for high precision flow measurement technology. We develop and produce flow meters for virtually all pumpable media and the corresponding evaluation electronics.whether for water or highly viscous adhesive media with filler material, our in-house development and design department produces technically sophisticated flow meters. The majority of our output is tailored to customer requirements, thereby putting handcraft back into industry. We can utilise modular systems that are adjusted in line with the technical specifications for each individual order. This means that product is carefully optimised for the needs of the customer. VSE flow meters are used primarily where it is critical to have the most accurate measurements possible. Our flow measurement technology

Range of Products

Equipment, plant and components
for conveying, mixing, metering and for adhesive application
for surface pretreatment
measuring and testing

For applications in the field of
Electronics
E-Mobility
Mechanical engineering and equipment construction
Chemical industry
Automotive industry, aviation industry
Textile industry
Hygiene
Household, recreation and office
Wood/furnitrure industry
Construction industry, including floors, walls and ceilings

is used worldwide in high-tech processing plants in the plastics, polyurethane, chemical, pharmaceutical, paint and varnish, hydraulic and automobile industries, as well as in 2-component technology.

Walther Spritz- und Lackiersysteme GmbH
Kärntner Straße 18 - 30
D-42327 Wuppertal
Phone +49 (0) 2 02-7870
Fax +49 (0) 2 02-7 87-22 17
Email: info@walther-pilot.de
www.walther-pilot.de

Company

Size of workforce
159 employees

Locations
Wuppertal-Vohwinkel
Neunkirchen-Struthütten

CEO
Ralf Mosbacher
Christian Glaser

Sales Manager
René Brettmann

Applications technology
Gerald Pöplau

Distribution channels
Field sales staff in Germany and
representatives in Europe and overseas.

Range of Products

Systems and components for the application
of adhesives, sealants and paints. Functioning
as a systems supplier, WALTHER engineers
customized, all-round solutions. They guarantee the best results over the long run in
terms of economy, user friendliness and
environment protection.

Application
Adhesive spray guns and automated
applicators
Extrusion guns
Metering valves
Spray guns for dots and lines
Multi-component metering and
mixing systems

Material conveyance
Pressure tanks
Diaphragm pumps
Piston pumps
Pump systems for high-viscosity media
Supply stations
Systems for shear-sensitive materials

Overspray exhaust
Spray booths
Filter technology
Ventilation systems

WALTHER PILOT

The Coating Experts

PILOT 2K BONDING
APPLICATION OF 2-COMPONENT ADHESIVE

ADVANTAGES AT A GLANCE:

- All parts that come into contact with material are made from stainless steel
- Easy to use
- High flexibility of use thanks to rotating wide jet air head
- All consumables can be quickly replaced
- Fine volume control for B-component

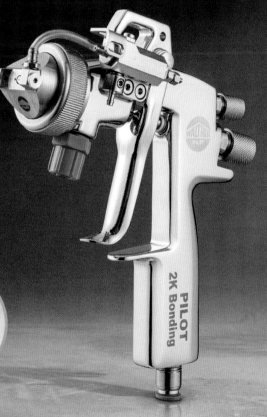

Robust stainless steel air head

Injection

www.walther-pilot.de

COMPANY PROFILES

Adhesive Technology
Consultancy Companies

ChemQuest Europe Inc.

Bilker Straße 27
D-40213 Düsseldorf
Phone +49 (0) 211-4 36 93 79
Fax +49 (0) 32 12-1 07 16 75
Mobile: +49 (0)1 71-3 41 38 38
Email: jwegner@chemquest.com
www.chemquest.com

Member of IVK

Company

Contact partners
Dr. Jürgen Wegner,
Managing Director
Email: jwegner@chemquest.com

Dr. Hubertus von Voithenberg,
Managing Director
Email: hvoithenberg@chemquest.com

CV and professional background
under www.chemquest.com

Consultancy

The ChemQuest Group is an international consulting firm headquartering in Cincinnati, Ohio, USA, with branch offices in Europe, China, North Africa and Latin America.

We specialize in consulting the Adhesives, Sealants, Construction Chemicals and Coatings Industry through all steps within the value chain from raw materials manufacturing through product formulation to all types of industrial and non-industrial end use applications. In South Boston/VA we recently acquired the ChemQuest Technology Institute, a fully equipped testing and formulating facility for the time being primarily focusing on coatings, supported by well experienced professional experts.

Based on in-depth knowledge our service portfolio includes all types of Management Consulting, M&A activities, market research, on coatings all types of quality and regulatory conformity tests, plus market and technology trend analysis. All associates of ChemQuest are experienced professionals from within the Adhesives, Sealants and Coatings Industry, and we are partnering with IFAM Fraunhofer Bremen in certified Adhesives education in North America. These courses currently take place at our South Boston facility. For further information and contact details please visit our website www.chemquest.com.

HINTERWALDNER CONSULTING

Our know-how – your future!

Consulting Chemists & Business Economists since 1956

Hinterwaldner Consulting
Dipl.-Kfm. Stephan Hinterwaldner
Markplatz 9
D-85614 Kirchseeon
Phone +49 (0) 80 91-53 99-0
Fax +49 (0) 80 91-53 99-20
Email: info@HiwaConsul.de
www.HiwaConsul.de

Member of IVK

Company

Year of formation
1956

Ownership structure
Privately held

Contact partner
Dipl.-Kfm. Stephan Hinterwaldner
Brigitte Schmid
Andrea Hinterwaldner

Adhesive Consulting
info@HiwaConsul.de

Adhesive Conferencing
contact@mkvs.de
contact@in-adhesives.com

Organizer/Co-organizer
- Munich Adhesives and Finishing Symposium
  Adhesives | Printing | Converting
  www.mkvs.de
- in-adhesives Symposium
  Adhesives | Industrial Adhesives Technology
  Automotive, Electronics, Medicine,
  Lightweight, Composites
  www.in-adhesives.com
- European Coatings Congress

Consultancy

Global Expert Research and Technology Consulting and Conferencing in the World of Adhesives

- Raw Materials, Ingredients, Intermediates, Additives
- Formulations, Applications, Product and Process Developments, Feasibility Studies
- Adhesives, Adhesive Tapes, Coatings, Cementations, Glues, Sealants, Release Liners
- Pressure Sensitive Adhesives, Hot Melt Adhesives, Chemically and Radiation Curing Adhesives Systems, Structural Bonding, Structural Glazing Systems
- Technologies in Adhesive Bonding, Coating, Converting, Film, Foil, Labeling, Laminating, Lightweight, Metalizing, Packaging, Printing, Sealing, Surface
- Polymer, Chemically and Radiation Curing, Petro-based, Bio-based and Green Chemistries
- Cosmetics and Toiletries, Beauty and Personal Care, Home Care
- Natural, Renewable, Sustainable, Bio-based and Certified Organic Products

Klebtechnik Dr. Hartwig Lohse e. K.

Hofberg 4
D-25597 Breitenberg
Phone +49 (0) 48 22-9 51 80
Fax +49 (0) 48 22-9 51 81
Email: hl@hdyg.de
www.how-do-you-glue.de

Member of the IVK

Company

Year of formation
2009

Contact partners
Dr. Hartwig Lohse

Further information
Based on 30+ years of industrial experience in the development, technical service and technical marketing of adhesives we are providing competent consultancy services along the adhesive supply chain. Our customers are industrial users of adhesives, adhesives manufacturing and raw material supply companies, equipment and plant manufacturing companies as well as manufacturers of dispensing and testing equipment.

Consultancy

Our consultancy service portfolio includes:

- optimisation of current and the planning of new bonding processes incl. generation of the specific requirements profile, selection of the best adhesive technology, source search for commercial adhesives, evaluation of adhesives, bond design advice, selection of suitable surface pre-treatment measures, Development and implementation of the bonding process into production (incl. adhesive dispense, surface pre-treatment, cure process, QC-measures, P-FMEA, etc.).
- implementation of DIN 2304 (Adhesive bonding technology - Quality requirements for adhesive bonding processes)
- adhesive failure analysis (troubleshooting)
- adhesive education, training at all levels specific to the clients requirements
- development of adhesives as well as raw materials for the use in manufacturing of adhesives
- market research, market & market trend analysis
- ...

For further information please visit us at www-how-do-you-glue.de

Contract Manufacturing and Filling Services

LOOP

LOHNFERTIGUNG UND OPTIMIERUNG

LOOP GmbH
Lohnfertigung und Optimierung
Am Nordturm 5
D-46562 Voerde
Phone +49 (0) 2 81-8 31 35
Fax +49 (0) 2 81-8 31 37
Email: mail@loop-gmbh.de

Member of IVK

Company

Year of formation
1993

Size of workforce
25

Contact partners
Managing Director:
DI Marc Zick

Production Manager:
Jürgen Stockmann

Further information
LOOP is consistently expanding its
personnel and equipment at plant and the
KST Customer Service Test Centre.

LOOP is partner of international and national well known companies.

Range of Products

Toll-Production of the following product groups
- polymer formulations, one- and two- component systems (unfilled, filled), aqueous, solvent
  based and solvent-free
- additives/concentrates of actives substances
  (liquid, pasty, powdery)
- impregnating and mould resin systems
- slurries
- powder mixtures
- substrates like SiO_2 coated with active
  substances
- Concentrate pastes
- Granulates and their fractions for applications
  in engineering and decorative applications
- compounds for electrical components
  and diverse additional product groups

LOOP operates on the following sectors for its partners
- adhesives and sealants
- polymer and binder chemicals
- pigments and fillers
- additives
- construction chemicals/preservation of
  buildings
- composites
- foundry products
- electronics
- cables
- power generation (wind power, photovoltaics)
- glass
- textile and paper finishes
- wood preservatives
- varnishes, paints, and printing inks
- professional development, laboratory, and
  consultancy services
- large scale distribution, etc.

COMPANY PROFILES

Research and Development

 **Fraunhofer**

IFAM

Fraunhofer Institute for Manufacturing Technology and Advanced Materials IFAM
– Adhesive Bonding Technology and Surfaces –

Wiener Straße 12
D-28359 Bremen
Phone +49 (0) 421 2246-400
Fax +49 (0) 421 2246-430
Email: info@ifam.fraunhofer.de
www.ifam.fraunhofer.de

Member of IVK

Company

Year of formation
1968

Size of workforce
All in all 700
Division of Adhesive Bonding Technology and Surfaces: > 400

Ownership structure
Fraunhofer IFAM is a constituent entity of the Fraunhofer-Gesellschaft zur Förderung der angewandten Forschung e. V. and as such has no separate legal status.

Contact
Fraunhofer IFAM
– Adhesive Bonding Technology and Surfaces –
Director: Prof. Dr. Bernd Mayer
Deputy director:
Prof. Dr. Andreas Hartwig

Work Areas

R&D – Contract-research and development – in all fields of adhesive bonding technology and surfaces as well as materials, in addition providing certifying training courses in adhesive bonding technology and fiber composite materials:

Adhesives and Polymer Chemistry:
Prof. Dr. Andreas Hartwig
Phone +49 (0) 421 2246-470
Email: andreas.hartwig@ifam.fraunhofer.de

Adhesive Bonding Technology:
Dr. Holger Fricke
Phone +49 (0) 421 2246-637
Email: holger.fricke@ifam.fraunhofer.de

Work Areas

Polymeric Materials and Mechanical Engineering:
Dr. Katharina Koschek
Phone +49 (0) 421 2246-698
Email: katharina koschek@ifam.fraunhofer.de

Automation and Production Technology (CFK Nord, Stade):
Dr. Dirk Niermann
Phone +49 (0) 4141 78707-101
Email: dirk.niermann@ifam.fraunhofer.de

Adhesion and Interface Research:
Dr. Stefan Dieckhoff
Phone +49 (0) 421 2246-469
Email: stefan.dieckhoff@ifam.fraunhofer.de

Plasma Technology and Surfaces PLATO:
Dr. Ralph Wilken
Phone +49 (0) 421 2246-448
Email: ralph.wilken@ifam.fraunhofer.de

Quality Assurance and Cyber-Physical Systems
Dipl.-Phys. Kai Brune
Phone +49 (0) 421 2246-459
Email: kai.brune@ifam.fraunhofer.de

Paint/Lacquer Technology:
Dr. Volkmar Stenzel
Phone +49 (0) 421 2246-407
Email: volkmar.stenzel@ifam.fraunhofer.de

Training Center for Adhesive Bonding Technology: Dr. Erik Meiß
Phone +49 (0) 421 2246-632
Email: erik.meiss@ifam.fraunhofer.de
www.bremen-bonding.com
Training Center for Fiber Composite Technology: Dipl.-Ing. Stefan Simon
Phone +49 (0) 421 5665-688
Email: stefan.simon@ifam.fraunhofer.de
www.bremen-composites.com

Berner
Fachhochschule

**Institute for Materials and
Wood Technology**
Solothurnstrasse 102, CH-2504 Biel
Phone +41 (0) 32 344 02 02
Fax +41 (0) 32 344 03 91
Email: iwh@bfh.ch
www.bfh.ch/de/forschung/forschungsbereiche/
institut-werkstoffe-holzechnologie-iwh/

Member of IVK

Company

Year of formation
1997

Size of workforce
2,424 employees

Managing partners
Bern University of Applied Sciences

Ownership structure
Governing Body: Canton of Bern

Contact partners
Frederic Pichelin
Email: frederic.pichelin@bfh.ch

Application technology and sales
Martin Lehmann
Email: martin.lehmann@bfh.ch

Further information
Focusing on the sustainable use of resources,
at the Institute for Materials and Wood
Technology of the Bern University of Applied
Sciences, we develop and optimise multifunctional wood and composite materials
and innovative products for the timber and
construction industries. These products
and materials are sourced from bio-based
materials such as wood and other renewable raw materials. We use our extensive
knowledge of these raw materials to help
us find innovative new uses for them.
In our work, we place particular emphasis
on process reliability and product quality.

Work Areas

Types of adhesives
Reactive adhesives
Dispersion adhesives
Vegetable adhesives, dextrin and starch
adhesives

Raw materials
Polymers

For applications in the field of
Wood/furniture industry
Construction industry, including floors, walls
and ceilings

SKZ - KFE GmbH

Friedrich-Bergius-Ring 22
D-97076 Würzburg
Phone +49 (0) 931 4104-0
Fax +49 (0) 931 4104-707
Email: kfe@skz.de
www.skz.de

Member of IVK

Company

Year of formation
1961

Size of workforce
204 in 2019

Managing partners
FSKZ e.V.

Nominal capital
35 TEUR

Ownership structure
100 % subsidiary

Contact partners
Management:
Mr. Dr. Thomas Hochrein

Application technology and sales:
Mr. Dr. Eduard Kraus

Further information
The German Plastics Center (SKZ) provides products from more than 400 companies. Its entities provide services in testing and certification of products and management systems, consulting, education, as well as research and development.

Work Areas

Adhesive Development
Additives
Fillers
Resins
Solvents

Application Development
Cleaning
Surface pretreatment
Adhesive application
Joining

Testing of adhesives and adhesive bonds

Trainings in adhesive technology

For applications in the field of
Adhesive Tapes and labels
Construction
Electronics
Mechanical engineering
Mobility industry (Automotive, Aviation, Naval, Railway)
Wood/furniture industry

ZHAW Laboratory of Adhesives

Laboratory of Adhesives and Polymer Materials
Zurich University of Applied Sciences (ZHAW)
Technikumstrasse 9, CH-8401 Winterthur
Phone +41 (0) 58 934 65 86
Email: christof.braendli@zhaw.ch
www.zhaw.ch/impe

Member of FKS

Company

Year of formation
School of Engineering: 1874, Institute: 2007

Size of workforce
School of Engineering: 580, Institute: 40

Managing partners
School of Engineering: Prof. Dr. Dirk Wilhelm,
Phone: +41 58 934 47 29,
Email: dirk.wilhelm@zhaw.ch
Institute of Materials and Process
Engineering: Prof. Dr. Andreas Amrein,
Phone: +41 58 934 73 51,
Email: andreas.amrein@zhaw.ch

Laboratory of Adhesives and Polymer
Materials. Prof. Dr. Christof Brändli,
Phone +41 58 934 65 86,
Email: christof.braendli@zhaw.ch

Ownership structure
Part of the Zurich University of Applied
Sciences (ZHAW)

Contact partners
Laboratory of Adhesives and Polymer
Materials: Prof. Dr. Christof Brändli,
Phone +41 58 934 65 86,
Email: christof.braendli@zhaw.ch

Further information
Applied R&D in the fields of materials,
adhesive bonding, and surfaces. Adhesive
development and testing. Broad range of
adhesive competencies including adhesive
chemistry and analysis.

Work Areas

Types of adhesives
Hot melt, 1K, 2K, UV, reactive, solvent-based, dispersion, pressure-sensitive
adhesives, sealants and resins

Top-down and bottom-up approaches
– Formulation, synthesis, application,
and analysis of adhesives & polymers
– Polymer compounding, polymer
modification/functionalization,
reactive extrusion, injection molding and
3D/SL printing
– Adhesive (structure-property relationship)
and polymer developments (incl. tapes,
film, grafting, reactive, etc.)

Analysis, Characterization and Processing
• Analytical infrastructure (SEM, AFM, opt.
3D microscopy, XPS, EDX, DSC, DMA,
rheology, NMR, tensile testing, SAFT, etc.)
Adhesive performance tests and curing
behavior studies

• Thermal and mechanical analysis
Flow properties determination
Morphological and surface analysis
Online reaction control with IR
spectroscopy
Unknown polymer identification

• Adhesive and polymer processing
Extruder & injection molding
Planetary, Speed mixer
3D/SL printing and 3D scanning
ABES (Automated Bonding Evaluation
System)
Laboratory hot melt coater

INSTITUTES AND RESEARCH FACILITIES

Research and Development

Adhesive bonding plays a significant role in the development of innovative products and provides all sectors of industry with the opportunity to open up new and promising markets. Small and medium-sized businesses can also make use of adhesive bonding to create ground-breaking products that will help them to remain competitive.

In order to take full advantage of the benefits of adhesive bonding, the entire process must be implemented correctly, from product planning and quality assurance through to employee training.

However, this is only possible when research institutions and industrial firms work closely together to ensure that the results of the research can be incorporated quickly and directly into the development of new products and production processes.

Below is a list of all the prominent research bodies and institutes that specialise in resolving adhesive bonding problems in a wide variety of areas with the cooperation of their industrial partners.

Deutsches Institut für Bautechnik (DIBt)
Abteilung II – Gesundheits- und Umweltschutz
(Department II – Health and Environmental
Protection)
Kolonnenstraße 30 B
D-10829 Berlin

Contact:
Dipl.-Ing. Dirk Brandenburger MEM (UTS)
Phone: +49 (0) 30 78730 232
Fax: +49 (0) 30 78730 11232
Email: dbr@dibt.de
www.dibt.de

Aachen University of Applied Sciences
Laboratory for Adhesive Technologies
Goethestraße 1
D-52064 Aachen

Contact:
Prof. Dr.-Ing. Markus Schleser
Phone: +49 (0) 241 6009 52385
Fax: +49 (0) 241 6009 52368
Email: schleser@fh-aachen.de
www.fh-aachen.de

Fogra Forschungsinstitut für
Medientechnologien e. V.
(Fogra Research Institute for
Media Technologies)
Einsteinring 1a
D-85609 Aschheim b. München

Contact:
Dr. Eduard Neufeld
Phone: +49 (0) 89 43182 112
Fax: +49 (0) 89 43182 100
Email: info@fogra.org
www.fogra.org

FOSTA Forschungsvereinigung
Stahlanwendung e. V.
(FOSTA Research Association for
Steel Application)
Steel Centre
Sohnstraße 65
D-40237 Düsseldorf

Contact:
Dipl.-Ing. Rainer Salomon
Phone: +49 (0) 211 30 29 76 00
Fax: +49 (0) 211 54 25 65 33
Email: rainer.salomon@stahlforschung.de
www.stahlforschung.de

Fraunhofer-Institut für Fertigungstechnik und
Angewandte Materialforschung – IFAM
(Fraunhofer Institute for Manufacturing
Technology and Advanced Materials IFAM)
Wiener Straße 12
D-28359 Bremen

Contact:
Prof. Dr. Bernd Mayer
Phone: +49 (0) 421 2246 419
Fax: +49 (0) 421 2246 774401
Email: bernd.mayer@ifam.fraunhofer.de
Prof. Dr. Andreas Groß
Phone: +49 (0) 421 2246 437
Fax: +49 (0) 421 2246 605
Email: andreas.gross@ifam.fraunhofer.de
www.ifam.fraunhofer.de

Fraunhofer-Institut für Holzforschung –
Wilhelm-Klauditz-Institut – WKI
(Fraunhofer Institute for Wood Research –
Wilhelm-Klauditz-Institute – WKI)
Bienroder Weg 54 E
D-38108 Braunschweig

Contact:
Dr. Heike Pecher
Phone: +49 (0)531 2155 206
Email: heike.pecher@wki.fraunhofer.de
www.wki.fraunhofer.de

Fraunhofer-Institut für Werkstoff- und
Strahltechnik – IWS
Fraunhofer Institute for Material and
Beam Technology – IWS
(Adhesives Technical Centre at Dresden
University of Technology, Institute of Manu-
facturing Science and Engineering,
Chair of Laser and Surface Technology)
Winterbergstraße 28
D-01277 Dresden

Contact:
Dr. Maurice Langer
Phone: +49 (0) 351 83391 3852
Fax: +49 (0) 351 83391 3210
Email: maurice.langer@iws.fraunhofer.de
www.iws.fraunhofer.de

Fraunhofer-Institut für Zerstörungsfreie
Prüfverfahren – IZFP
(Fraunhofer Institute for Non-Destructive
Testing IZFP)
Campus E3.1
D-66123 Saarbrücken

Contact:
Prof. Dr.-Ing. Bernd Valeske
Phone: +49 (0) 681 9302 3610
Fax: +49 (0) 681 9302 11 3610
Email: bernd.valeske@izfp.fraunhofer.de
www.izfp.fraunhofer.de

Johann Heinrich von Thünen Institut (vTI)
Bundesforschungsistitut für Ländliche Räume,
Wald und Fischerei
Institut für Holzforschung
(The Thünen Institute
German Federal Research Institute for Rural
Areas, Forestry and Fisheries
Institute of Wood Research)
Leuschnerstraße 91
D-21031 Hamburg

Contact:
Prof. Andreas Krause
Phone: +49 (0) 40 73962 623
Fax: +49 (0) 40 73962 699
Email: andreas.krause@thuenen.de
www.thuenen.de/de/hf/

Hochschule München
Institut für Verfahrenstechnik Papier e.V. (IVP)
(Munich University of Applied Sciences
Institute of Paper Technology (IVP))
Schlederloh 15
D-82057 Icking

Contact:
Prof. Dirk Burth
Phone: +49 (0) 89 1265 1668
Fax: +49 (0) 89 1265 1560
Email: info@ivp.org
https://ivp.org/de

Hochschule für nachhaltige Entwicklung
Eberswalde (FH)
Fachbereich Holzingenieurwesen
(Eberswalde University for Sustainable
Development – Faculty of Wood Engineering)
Schlickerstraße 5
D-16225 Eberswalde

Contact:
Prof. Dr.-Ing. Ulrich Schwarz
Phone: +49 (0) 3334 657 371
Fax: +49 (0) 3334 657 372
Email: ulrich.schwarz@hnee.de
www.hnee.de/holzingenieurwesen

IFF GmbH
Induktion, Fügetechnik, Fertigungstechnik
(Induction, Bonding and Manufacturing
Technology)
Gutenbergstraße 6
D-85737 Ismaning

Contact:
Prof. Dr.-Ing. Christian Lammel
Phone: +49 (0) 89 9699 890
Fax: +49 (0) 89 9699 8929
Email: christian.lammel@iff-gmbh.de
www.iff-gmbh.de

ift Rosenheim GmbH
Institut für Fenstertechnik e.V.
(Institute for Window Technology)
Theodor-Gietl-Straße 7-9
D-83026 Rosenheim

Contact:
Prof. Peter Lass
Phone: +49 (0) 80 31261 0
Fax: +49 (0) 80 31261 290
Email: info@ift-rosenheim.de
www.ift-rosenheim.de

ihd – Institut für Holztechnologie Dresden GmbH
(Institute of Wood Technology)
Zellescher Weg 24
D-01217 Dresden

Contact:
Dr. rer. nat. Steffen Tobisch
Phone: +49 (0) 351 4662 257
Fax: +49 (0) 351 4662 211
Mobil: +49 (0) 1622 696330
Email: steffen.tobisch@ihd-dresden.de
www.ihd-dresden.de

Georg-August-Universität Göttingen
Institut für Holzbiologie und Holzprodukte
(Institute for Wood Biology and Wood
Technology)
Büsgenweg 4
D-37077 Göttingen

Contact:
Prof. Dr. Holger Militz
Phone: +49 (0) 551 3933541
Fax: +49 (0) 551 399646
Email: hmilitz@gwdg.de
www.uni-goettingen.de

Institut für Fertigungstechnik
Professur für Fügetechnik und Montage
TU Dresden
(Institute of Manufacturing Science and
ngineering
Chair of Joining Technology and Assembly
Dresden University of Technology)
George-Bähr-Straße 3c
D-01069 Dresden

Contact:
Prof. Dr.-Ing. habil. Uwe Füssel
Phone: +49 (0) 351 46337 615
Fax: +49 (0) 351 46337 249
Email: uwe.fuessel@tu-dresden.de
https://tu-dresden.de/ing/
maschinenwesen/if/fue

IVLV - Industrievereinigung für Lebens-
mitteltechnologie und Verpackung e.V.
(IVLV – Industry Association for Food
Technology and Packaging)
Giggenhauser Straße 35
D-85354 Freising

Contact:
Dr.-Ing. Tobias Voigt
Phone: +49 (0) 8161 491140
Fax: +49 (0) 8161 491142
Email: tobias.voigt@ivlv.org
www.ivlv.org

iwb - Institut für Werkzeugmaschinen und
Betriebswissenschaften
Fakultät für Maschinenwesen - Themengruppe
Füge- und Trenntechnik
Technische Universität München
(iwb – Faculty of Mechanical Engineering –
Department of Laser Technologies
Technical University of Munich)
Boltzmannstraße 15
D-85748 Garching b. München

Contact:
Stefan Meyer, M.Sc.
Phone: +49 (0) 89 289 15548
Email: stefan.meyer@iwb.tum.de
www.mw.tum.de/iwb

Kompetenzzentrum Werkstoffe der Mikrotechnik
Universität Ulm
(Centre of Excellence of Micro and Nano
Materials University of Ulm)
Albert-Einstein-Allee 47
D-89081 Ulm

Contact:
Prof. Dr. Hans-Jörg Fecht
Phone: +49 (0) 731 50254 91
Fax: +49 (0) 731 50254 88
Email: info@wmtech.de
www.wmtech.de

Leibniz-Institut für Polymerforschung
Dresden e.V.
(Leibniz Institute of Polymer Research Dresden)
Hohe Straße 6
D-01069 Dresden

Contact:
Prof. Dr.-Ing. Udo Wagenknecht
Phone: +49 (0) 351 46583 61
Fax: +49 (0) 351 46583 62
Email: wagenknt@ipfdd.de
www.ipfdd.de

Naturwissenschaftliches und Medizinisches
Institut an der Universität Tübingen
(Natural and Medical Sciences Institute
University of Tübingen)
Markwiesenstraße 55
D-72770 Reutlingen

Contact:
Dr. Hanna Hartmann
Phone: +49 (0)7121 51530 872
Fax: +49 (0)7121 51530 62
Email: hanna.hartmann@nmi.de
www.nmi.de

ofi Österreichisches Forschungsinstitut
für Chemie und Technik
Technische Kunststoffbauteile, Klebungen &
Beschichtungen
(Austrian Research Institute for Chemistry
and Technology (ofi)
Industrial plastic components, adhesives
and coatings)
Franz-Grill-Straße 1, Objekt 207
A-1030 Wien

Contact:
Ing. Martin Tonnhofer
Phone: +43 (0)1798 16 01 201
Email: martin.tonnhofer@ofi.at
www.ofi.at

Papiertechnische Stiftung PTS
(Paper Technology Foundation)
Pirnaer Straße 37
D-01809 Heidenau

Contact:
Clemens Zotlöterer
Phone: +49 (0) 3529 551 60
Fax: +49 (0) 3529 551 899
Email: info@ptspaper.de
www.ptspaper.de

Prüf- und Forschungsinstitut Pirmasens e.V.
(Test and Research Institute Pirmasens)
Marie-Curie-Straße 19
D-66953 Pirmasens

Contact:
Dr. Kerstin Schulte
Phone: +49 (0)6331 2490 0
Fax: +49 (0)6331 2490 60
Email: info@pfi-germany.de
www.pfi-germany.de

RWTH Aachen
ISF - Institut für Schweißtechnik und
Fügetechnik
(RWTH Aachen University
Welding and Joining Institute (ISF))
Pontstraße 49
D-52062 Aachen

Contact:
Prof. Dr. -Ing. Uwe Reisgen
Phone: +49 (0) 241 80 93870
Fax: +49 (0) 241 80 92170
Email: office@isf.rwth-aachen.de
www.isf.rwth-aachen.de

Technische Universität Berlin
Fügetechnik und Beschichtungstechnik im
Institut für Werkzeugmaschinen und
Fabrikbetrieb
(Berlin Institute of Technology
Joining and Coating Technology at the Institute of
Machine Tools and Manufacturing Technology)
Pascalstraße 8 – 9
D-10587 Berlin

Contact:
Prof. Dr.-Ing. habil. Christian Rupprecht
Phone: +49 (0) 30 314 25176
Email: info@fbt.tu-berlin.de
www.fbt.tu-berlin.de

Technische Universität Braunschweig
Institut für Füge- und Schweißtechnik
(Technical University Braunschweig
Institute of Joining and Welding Technology)
Langer Kamp 8
D-38106 Braunschweig

Contact:
Univ.-Prof. Dr.-Ing. Prof. h.c. Klaus Dilger
Phone: +49 (0) 531 391 95500
Fax: +49 (0) 531 391 95599
Email: k.dilger@tu-braunschweig.de
www.ifs.tu-braunschweig.de

Technische Universität Kaiserslautern
Fachbereich Maschinenbau und
Verfahrenstechnik
Arbeitsgruppe Werkstoff- und Oberflächen-
technik Kaiserslautern (AWOK)
(Technical University of Kaiserslautern
Department of Mechanical and Process
Engineering Workgroup Materials and Surface
Technologies (AWOK))
Building 58, Room 462
Erwin-Schrödinger-Straße
D-67663 Kaiserslautern

Contact:
Univ.-Prof. Dr.-Ing. Paul Ludwig Geiß
Phone: +49 (0) 631 205 4117
Fax: +49 (0) 631 205 3908
Email: geiss@mv.uni-kl.de
www.mv.uni-kl.de/awok

TechnologieCentrum Kleben
TC-Kleben GmbH
(Center Adhesive Bonding Technology)
Carlstraße 54
D-52531 Übach-Palenberg

Contact:
Dipl.-Ing. Julian Band
Phone: +49 (0) 2451 9712 00
Fax: +49 (0) 2451 9712 10
Email: post@tc-kleben.de
www.tc-kleben.de

Universität Kassel
Institut für Werkstofftechnik, Kunststofffüge-
techniken, Werkstoffverbunde
(University of Kassel
Institute of Materials Engineering, Plastics
Joining Technology, Materials Composites)
Mönchebergerstraße 3
D-34125 Kassel

Contact:
Prof. Dr.-Ing. Hans-Peter Heim
Phone: +49 (0) 561 80436 70
Fax: +49 (0) 561 80436 72
Email: heim@uni-kassel.de
www.uni-kassel.de/maschinenbau

Universität Kassel
Fachgebiet Trennende und
Fügende Fertigungsverfahren
(University of Kassel
Department of Cutting and Joining
Manufacturing Processes)
Kurt-Wolters-Straße 3
D-34125 Kassel

Contact:
Prof. Dr.-Ing. Prof. h.c. Stefan Böhm
Phone: +49 (0) 561804 3236
Fax: +49 (0) 561804 2045
Email: s.boehm@uni-kassel.de
www.tff-kassel.de

Universität des Saarlandes
Adhäsion und Interphasen in Polymeren
(University of Saarland
Chair for Adhesion and Interphases in Polymers)
Campus, Building C6.3
D-66123 Saarbrücken

Contact:
Prof. Dr. rer. nat. Wulff Possart
Phone: +49 (0) 681 302 3761
Fax: +49 (0) 681 302 4960
Email: w.possart@mx.uni-saarland.de
www.uni-saarland.de/fak8/wwthd/
index.ger.html

Universität Paderborn
Laboratorium für Werkstoff- und Fügetechnik
(Laboratory of Materials and Joining
Technology (LWF))
Pohlweg 47-49
D-33098 Paderborn

Contact:
Prof. Dr.-Ing. Gerson Meschut
Phone: +49 (0) 5251 603031
Fax: +49 (0) 5251 603239
Email: meschut@lwf.upb.de
www.lwf-paderborn.de

Wehrwissenschaftliches Institut für Werk- und
Betriebsstoffe - WIWeB
(Bundeswehr Research Institute for Materials,
Fuels and Lubricants – WIWeB)
Institutsweg 1
D-85435 Erding

Contact:
Phone: +49 (0) 81 229590 0
Fax: +49 (0) 81 229590 3902
Email: wiweb@bundeswehr.org
www.bundeswehr.de/de/organisation/
ausruestung-baainbw/organisation/wiweb

Westfälische Hochschule Abteilung
Recklinghausen
Fachbereich Ingenieur- und
Naturwissenschaften
Organische Chemie und Polymere
(Westphalian University of Applied Sciences,
Recklinghausen Campus
Department of Engineering and Science
Organic Chemistry and Polymers)
August-Schmidt-Ring 10
D-45665 Recklinghausen

Contact:
Prof. Dr. Klaus-Uwe Koch
Phone: +49 (0) 2361 915 456
Fax: +49 (0) 2361 915 751
Email: klaus-uwe.koch@w-hs.de
www.w-hs.de/erkunden/fachbereiche/
ingenieur-und-naturwissenschaften/
portrait-des-fachbereichs/

*German
Adhesives
Association*
Industrieverband Klebstoffe e.V.

ANNUAL REPORT 2020/2021

Economic Report

Since the start of the COVID-19 pandemic in early 2020, the German adhesives industry has been on a roller coaster ride. After a good to very good first quarter of 2020, the industry was hit in March 2020 by the first wave of infection, which led to serious losses in Q2 when compared with the previous year. Almost all market segments were affected, but in very different ways. Manufacturers of products for the automotive industry and its suppliers felt the full and direct impact of the immediate shutdown of this market segment. However, suppliers of adhesive systems to sectors such as (food) packaging, hygiene products, medicines, medical technology, pharmaceuticals and electronics saw their sales remain largely stable. Adhesives intended for trade users and for consumer goods were increasingly affected by consumers' unwillingness to spend. The entire industry experienced the first signs of recovery in Q3. In Q4, sales reached record levels because of the catch-up effect. While the very good order situation in early 2021 had a highly positive impact on the economic recovery, the combination of the unexpectedly high increase in orders and the number of manufacturers in Europe and the USA invoking force majeure clauses had a direct and long-term negative effect on the availability of important adhesive raw materials. In addition, the pandemic continued to exert a negative influence on business activities. The industry's domestic revenues in 2020 from all types of adhesive systems, including adhesives, sealants, cement-based systems and adhesive tapes, amounted to € 3.85 billion.

The economic situation in 2020 – no forecast possible for 2021

The economic environment remained unpredictable and volatile for the German adhesives industry throughout 2020 and into 2021. The COVID-19 pandemic and the limited availability of adhesives are also continuing to have an impact on the economy. Against the background of the current fragile and unstable socioeconomic situation, it is not possible to make anything resembling a reliable forecast of the economic growth of the German adhesives industry for 2021. After the economic recovery, it is expected that the influence of megatrends such as urbanisation, electric mobility, autonomous driving and the Internet of Things will create a much more positive market environment for the German adhesives industry.

COVID-19 pandemic – economic effects

The COVID-19 pandemic is not only a global health crisis; it will also have a significant negative impact on the global economy, with consequences that cannot as yet be foreseen. The German economy has experienced a sharp recession, but some market segments that are important for the adhesives industry (packaging, construction and electronics) have remained amazingly robust. In March and April 2020, there was a record collapse in automotive production, for example, but some signs of a slow recovery could be seen at the end of the year.

In Q2 and Q3 2021, the difficult situation on the raw materials market is having a negative influence on the economic recovery.

The German adhesives industry on competitive European and international markets

Disregarding for a moment the possible economic impacts of the COVID-19 pandemic, the German adhesives industry is in a very strong position on European and other international markets. With a global market share of more than 19 percent, it is the world market leader. It also holds the top positions on the European market, with an adhesives consumption of 27 percent and a production share of more than 34 percent.

Worldwide, sales revenues of approximately € 61 billion per year (not adjusted for exchange rate effects) were generated with adhesives, sealants, adhesive tapes and system products. The German adhesives industry, which consists mainly of medium-sized companies, plays a prominent role on the international stage. Most of the companies manufacture their products in Germany and export them worldwide. In addition, some German adhesives firms supply the world markets from more than 200 local manufacturing plants outside Germany.

With both business models, the German adhesives industry generates sales revenues of almost € 12 billion worldwide. Exports from Germany amount to more than € 1.7 billion, while German adhesives manufacturers generate a further € 8.1 billion of sales revenues locally from their production facilities outside Germany. The German market has a sales volume of almost € 4 billion per year. As a result of the use of "adhesive systems developed in Germany" in almost all manufacturing industries and in the construction sector, the adhesives industry was responsible for indirect value creation of significantly more than € 400 billion in Germany. Worldwide, value creation amounted to more than € 1 trillion.

This strong position is the direct result of innovative technological developments for all the key market segments and applications where the adhesives industry provides practical, value-added solutions.

The link between the price of crude oil and of basic raw materials for adhesives weakens

The price of crude oil is no longer directly reflected in the price of end products. This connection has almost been broken. Prices are increasingly being determined by the many refining processes involved in the value chain (see the graphic "Raw material flows") and the availability of a raw material on the market. The link between the price of a range of basic raw materials and that of crude oil is likely to weaken even further. This is made very clear by a comparison between the fluctuations in the price of crude oil and of acetic acid, VAM, ethylene and methanol (see the graphic "Example of price fluctuations in crude oil and basic raw materials"). While the price of crude oil fell throughout 2015 and 2016 for example, the prices of the raw materials referred to above remained largely stable or were heavily influenced by regional supply and demand. In particular in the case of high-quality adhesives, the fluctuations in prices are no longer linked because there can be up to 10 processing stages between crude oil and the raw materials used to manufacture PUR, epoxy resin and acrylate adhesives.

Increases in the prices of basic chemicals and intermediate products combined with limited availability and a strong global demand for a variety of adhesive raw materials is and will remain

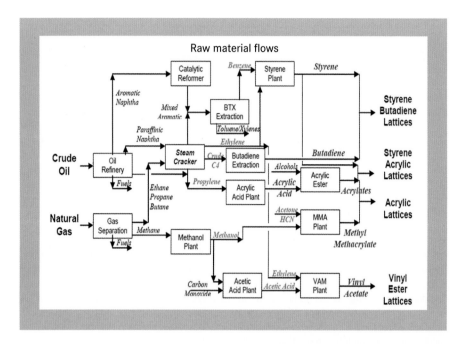

Raw material flows

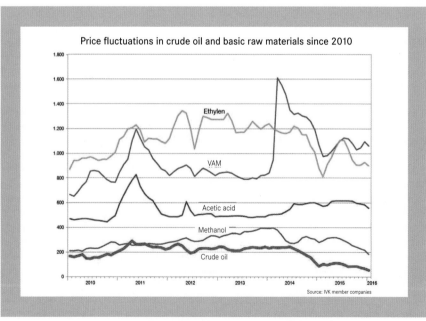

Price fluctuations in crude oil and basic raw materials since 2010

Source: IVK member companies

a long-term issue for the adhesives industry, which will vary in intensity. In addition, the constantly rising costs imposed by regulations and a shortage of freight capacity in Europe resulted in a noticeable increase in pressure on the cost structure of the German adhesives industry.

About our Committee Work

General Assembly

The annual general meeting and general assembly due to be held in 2020 in Leipzig had to be cancelled because of the COVID-19 pandemic. To ensure that the members and committees received the information they needed, for the first time the association provided written reports about its work, the market situation and the latest technical and legislative developments. The votes in the elections to the association's committees, which were planned for 2020, were also submitted in writing under the terms of the Act to Mitigate the Consequences of the COVID-19 Pandemic. To guarantee the secrecy of the ballot in accordance with the association's statutes, the votes were counted by an agency.

The 2021 general assembly was held on 28 May 2021 for the first time in online form. A total of 186 representatives of 81 member companies took part in the event.

Economic growth

Overall, the adhesives industry has come out of the COVID-19 crisis relatively intact. After satisfactory growth in the market for adhesives in the first two months of 2020, the German adhesives industry was hit by the global pandemic in March 2020, but the extent of the impact varied across the different key markets. While manufacturers of products for the automotive industry and its suppliers felt the full and direct impact of the immediate shutdown of this market segment, suppliers of adhesive systems to sectors such as (food) packaging, hygiene products, medicines, medical technology, pharmaceuticals and electronics saw their sales remain largely stable. The manufacturers of adhesives for the timber and furniture industries benefited from the cocooning effect.

The volume of adhesives produced in Germany in 2020 was 3.7 percent lower than in 2018 and the production value dropped by 5.1 percent. Imports fell by 8 percent and exports by 3 percent over the previous year.

It was expected that the influence of megatrends in 2020 and the following years would help to create a much more positive market environment and that this would be reflected in the economic figures (GDP, IPX).

The results of the IVK market sentiment survey show that the majority of companies take a positive view of the future. Demand is no longer a cause for anxiety, but rises in raw material prices and supply bottlenecks are giving rise to concern.

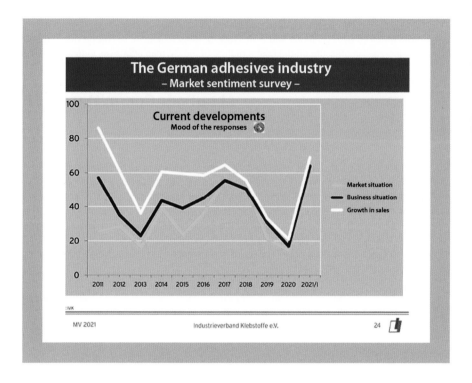

The work of the association

The services that the German Adhesives Association (IVK) provides for its members cover three key areas:
• Technology
• Communication
• Future initiatives

IVK continues to give the highest priority to technical subjects because working together to manage and implement the legislative requirements in particular that apply to adhesives companies has a significant impact on the success of the industry. This applies to REACH, labelling (GHS/CLP), permitted biocides, (mandated) standardisation and sustainability, as well as technical briefing notes, seminars and conferences and contacts with customer organisations.

The specialist groups:
• Adhesives for electronics
and
• Adhesives for textiles

are two new committees created following a resolution by the Executive Board and the general assembly because electronics and textiles form an important part of the EU's Green Deal and Circular Economy Action Plan.

The involvement of the German Adhesives Association in standards activities highlights the leading role it plays on a technical level in Europe and throughout the rest of the world. The IVK standardisation competence platform has enabled important German industry standards to be established and implemented. These are increasingly being followed as the markets for adhesives become more globalised.

The association's communications and public relations activities remain extremely important. In the summer of 2020, the IVK website was completely revised and is now more modern and easier to navigate. The content has been restructured to make access to the information more intuitive and faster. In addition, the IVK online magazine "Kleben fürs Leben" (Bonding for Life) has been coordinated with and integrated into the website. The site has been designed for all the current browsers and mobile devices, including tablets, and meets the latest standards. Information and downloads are still available exclusively to members of the association in a separate intranet area.

This year's print edition of "Kleben fürs Leben" (Bonding for Life), created in cooperation with the adhesives associations of Austria and Switzerland, was sent out to coincide with the virtual annual general meeting. This anniversary issue had a cover story entitled "The History of Adhesives/75 Years of IVK". As usual, it contains an interesting mix of articles on science, lifestyle matters and everyday concerns.

The 03/2021 issue of "FussbodenTechnik" magazine, the trade publication for contractors, floor layers and screeders, contains a 55-page special section on 75 years of the German Adhesives Association, with contributions from employees of the IVK office and representatives of several member companies.

IVK also has a presence on the relevant social media platforms, including Facebook, LinkedIn, Twitter and YouTube. Adhesive bonding is a theme across all the social media channels.

The educational programme "Die Kunst des Klebens" (The Art of Bonding) has been digitized and linked with other information material provided by the association (the e-paper "Job Profiles", the promotional video, the adhesives guidelines, the adhesives magazine etc.). As a result, since the start of the 2018/2019 school year, teachers have had a tool at their disposal that meets the new requirements for digital vocational education. The digital programme was downloaded by many teachers during the first phase of the COVID-19 pandemic in particular, when all the schools were closed and teaching moved online. It was also used by parents for home schooling. The materials are still available to download free of charge: www.klebstoffe.com/informationen/unterrichtsmaterialien/.

In November 2019, IVK collaborated with the North Rhine-Westphalia branch of the German Chemical Industry Association (VCI) to organise advanced training for teachers, which was financed by the German Chemical Industry Fund (FCI). A practical and theoretical one-day training session was held at three different IVK member companies. Around 50 people took part and went away from the course with a box containing all the adhesives necessary for the tests, so that they could carry out the tests in their lessons. The advanced teacher training for 2020/2021 unfortunately had to be cancelled because of the COVID-19 pandemic.

The association's Technical Committees have produced more than 60 publications (technical briefing notes, reports, information sheets). These are an important and much valued source of information for the customers of the adhesives industry. Almost all of them are also available in English.

The subject of sustainability is becoming increasingly important. On a political level, the European Commission's Green Deal involves a wide range of ambitious measures to protect the climate, all of which are aimed at making Europe the first climate-neutral continent by 2050. In order to provide expert technical support in all these areas in future, IVK established an advisory board for sustainability in 2018. At the suggestion of the Executive Board and with the approval of the general assembly, IVK commissioned the Fraunhofer Institute for Manufacturing Technology and Advanced Materials IFAM to carry out a scientific study entitled "Circular Economy and Adhesive Bonding Technology" which was completed in mid 2020. The study was supported by the Technical Board and the advisory board for sustainability. The study is available to download free of charge: https://www.ifam.fraunhofer.de/de/Presse/Kreislaufwirtschaft-Klebtechnik. html (German version "Kreislaufwirtschaft und Klebtechnik") and https://www.ifam.fraunhofer.de/ en/Press_Releases/adhesive_bonding_circular_economy.html (English version: "Circular Economy and Adhesive Bonding Technology").

On 25 November 2020, the study was presented to interested parties including consumers, researchers, politicians and representatives of the public sector during an online seminar held by the Fraunhofer IFAM. The institute gave an objective and easily understandable explanation of why adhesive bonding can be beneficial in a circular economy.

The adhesives industry believes that sustainability represents a new regulatory area which will have a major influence on the requirements placed on adhesive bonding in the decades to come.

The sample EPDs (sample environmental product declarations for construction adhesives) initiated by the German Adhesives Association and the accompanying guidelines are available on the IVK website at https://www.klebstoffe.com/nachhaltigkeit/.

The original national guidelines have in the meantime been made into a European standard, certified by IBU and published on the FEICA Internet portal as European sample EPDs. These documents are widely recognised in Europe, but in some cases additional national data is required.

In addition, IVK and its committees are actively working on a number of different issues and projects as part of the association's portfolio of services, which is based on the needs of its members. These include:

- The introduction of the DIN 2304 standard "Quality requirements for adhesive bonding processes"
- Support by EMICODE for very low emission products
- The approval of in-can preservatives and changes in labelling requirements
- REACH restriction on diisocyanates
- Reports to national poison centres
- REACH and REACH for polymers
- Microplastics
- ...

As the official production statistics from the German Federal Statistical Office no longer provide reliable information, the association has drawn up production statistics internally which have no implications with regard to competition law and involve very little work for the member companies. This survey of member companies, which is carried out on an annual basis by an agency, aims to make it possible to document the economic importance of the German adhesives industry on competitive European and international markets.

Since the 2019 general assembly, the following companies have been accepted as members of the German Adhesives Association:
- ALFA Klebstoffe AG
- Arakawa Europe GmbH
- Dunlop Tech GmbH
- ekp coatings GmbH
- Gustav Grolman GmbH & Co. KG
- HANSETACK GmbH
- H&H Maschinenbau GmbH
- INPROPAK GmbH
- Kissel + Wolf GmbH
- MORCHEM GmbH
- Poly-clip System GmbH & Co. KG
- Reka Klebetechnik GmbH & Co. KG

IVK now has 145 members, of which 122 are ordinary members, 11 are special members and 12 are associate members.

Sustainability

The concept of sustainability influences what we do in all areas of our lives, including the use of adhesives. Sustainability involves developing future-proof, long-lasting products that will not only meet our immediate needs, but will also satisfy the requirements of future generations, while taking into account environmental, economic and social factors.

The European Commission's Green Deal involves a wide range of ambitious measures to protect the climate, all of which are aimed at making Europe the first climate-neutral continent by 2050. The Circular Economy Action Plan and one of its key instruments, the Ecodesign Directive, are currently having an impact on the adhesives industry. The Ecodesign Directive creates a framework for establishing ecodesign requirements for energy-related products. Alongside energy efficiency, the important considerations for the European Commission include resource and material efficiency, durability, repairability and reusability. The Renovation Wave, the EU Industrial Strategy and the Chemicals Strategy for Sustainability are among other important measures that form part of the Green Deal and present both challenges and opportunities for the adhesives industry. On the one hand, energy-saving lightweight structures are inconceivable without the use of adhesives, while, on the other hand, adhesives are suspected of being an obstacle to the circular economy. The EU believes that there is a conflict of objectives here between energy efficiency and resource efficiency. Therefore, in discussions with public bodies, in a variety of public consultations and in a position paper, IVK has made it clear that:

• Bonding does not prevent a product from being repaired or recycled. (However, debonding must be one of the requirements placed on product manufacturers and must be taken into consideration during the product design process, in collaboration with the adhesives supplier.)

• Only a formulation which is technology-neutral (the specification of objectives instead of joining methods) will allow innovations to be developed in future.

The responsible use of resources is also a fundamental requirement of the EU Construction Products Regulation, which concerns buildings and therefore construction adhesives. The construction and building industry generates a significant proportion of our greenhouse gas emissions and uses large amounts of energy. One prerequisite for sustainable development is an accurate life cycle assessment covering the environmental impacts of the construction products being used. Environmental product declarations (EPDs) are a standardised method of describing the life cycle assessment of a construction product. Adhesives with a variety of formulations are used for a wide range of applications in buildings and this makes the creation of individual, product-specific EPDs prohibitive for cost reasons. Therefore, a system of sample EPDs has been developed which cover specific formulations and applications and which allow the effects of the products with the greatest environmental impact to be specified. The sample EPDs can be used by all IVK members: https://www.klebstoffe.com/epd-nachhaltigkeit/. In addition to measures and strategies on a European and national level, it is also important to monitor the initiatives launched by a variety of companies, industries and organisations that use adhesives.

In order to provide expert technical support in all these areas in future, IVK established an advisory board for sustainability at the end of 2018, which meets three times a year. At the suggestion of the Executive Board and with the approval of the general assembly, IVK commissioned the Fraunhofer Institute for Manufacturing Technology and Advanced Materials IFAM to carry out a scientific study entitled "Circular Economy and Adhesive Bonding Technology" which was completed in mid 2020. The study was supported by the Technical Board and the advisory board for sustainability. It is available to download free of charge: https://www.ifam. fraunhofer.de/de/Presse/Kreislaufwirtschaft-Klebtechnik.html (German version) and https:// www.ifam.fraunhofer.de/en/Press_Releases/adhesive_bonding_circular_economy.html (English version).

The IVK Executive Board believes that sustainability represents a new regulatory area which will have a major influence on the requirements placed on adhesive bonding in the decades to come.

Guidelines "Adhesive Bonding – the Right Way"

Creating a strong bonded joint involves more than just choosing the right adhesive. Other important factors that need to be taken into consideration include the properties of the materials, the surface treatment, a design suitable for bonding and proof of the reliability of the joint in use. The German Adhesives Association has collaborated with the Fraunhofer Institute IFAM to develop the interactive guidelines "Adhesive bonding – the right way". They are intended for tradespeople and industrial businesses that need additional information about adhesive systems.

Adhesives now have such a wide variety of uses that it is not possible for adhesives manufacturers to cover all the possible applications and, in particular, highly specialised ones in their data sheets. The guidelines "Adhesive bonding – the right way", which have been created by the German Adhesives Association and the Fraunhofer Institute IFAM, are a practical guide for tradespeople and industrial companies that need basic or more advanced information. The process of designing, developing and manufacturing an imaginary product is explained step by step and all the stages in the planning and manufacturing process are covered systematically. Carrying out all the necessary steps in the correct order makes a significant contribution to the quality of the end product. The interactive guidelines are the ideal tool for achieving this and they also include a glossary and search function. They cover the most important areas in the practical use of adhesives technology.

The German version of the guidelines is available free of charge in interactive form on the web portal of the German Adhesives Association: http://leitfaden.klebstoffe.com..

The English version of the guidelines is available free of charge in interactive form on the web portal of the German Adhesives Association: http://onlineguide.klebstoffe.com.

Executive Board

The membership of the Executive Board of the German Adhesives Association reflects the existing corporate structure of the German adhesives industry, which consists of small and medium-sized businesses as well as companies operating at a multinational level. Furthermore, the balance in the board's members guarantees access to the highest level of core competence and expertise relating to the key market segments that are important to the adhesives industry.

A top priority for the Executive Board of the German Adhesives Association is to adapt the association's structure and its committees to new political, economic and technological constraints and conditions quickly and on an ongoing basis, in order to ensure that the organisation always operates efficiently and provides the maximum benefit for the German adhesives industry.

Holding technical discussions, assessing the economic, political and technological trends in the various key market segments of the adhesives industry and monitoring and analysing the activities of the association's numerous commercial and technical committees form an integral part of the responsibilities of the association's Executive Board.

The German Adhesives Association is considered to be the foremost competence centre in the field of adhesive bonding and sealing. It is the world's largest national association and the leading body in terms of its comprehensive service portfolio for adhesive bonding technology. IVK's links with the German Chemical Industry Association (VCI) and its various technical sections form the basis for its successful position. In addition to its connections with the chemicals industry, the association has a strategic and highly-efficient 360-degree network of expertise consisting of all the relevant system partners, scientific institutions, leading trade and industry associations, employers' liability insurance associations, consumer organisations and the organisers of exhibitions, training courses and conventions. As a result, it covers every single element in the adhesives value chain.

Within the framework of the strategy developed by the association's Executive Board for a qualified market expansion, IVK actively supports a wide variety of scientific research projects relating to the field of adhesive bonding technology. This systematic approach to research is primarily interdisciplinary. In practical terms, this means that scientific knowledge is combined with engineering expertise in order to produce practical research results. Consequently, this approach has contributed to the fact that adhesive bonding has now become predictable and is firmly established as a reliable bonding and sealing technology in the field of engineering. It is now rightly and undisputedly considered to be the key technology of the 21st century.

As a founding member of the ProcessNet adhesives group under the umbrella of DECHEMA and the joint committee on adhesive bonding (GAK), IVK maintains regular contact with all the relevant research institutions involved in steel, timber and automotive research. Together, they jointly assess and support publicly funded scientific research projects in the field of adhesive bonding technology. This cooperation has enabled the German Adhesives Association to successfully establish important research projects with results that promise significant benefits in particular for the adhesives companies which are members of the association. The approved project concepts and the associated results are regularly presented during the annual DECHEMA colloquium "Gemeinsame Forschung in der Klebtechnik" (Joint research in adhesives and bonding technology).

The employee training programme of the Fraunhofer IFAM in Bremen, which has received strong support from IVK in terms of finance and content, has evolved into a well-established, recognised educational programme. More and more companies that use adhesives have realised the measurable advantages of providing their employees with training in the appropriate use of technically demanding adhesive systems. For example, bonded joints used in the manufacture of rail vehicles are now produced only by trained staff. The adhesives industry itself benefits in every respect from expert system partners.

To supplement the employee training programme, a DIN standard that ensures the quality of load-bearing/structural bonded joints used in specific safety areas has been developed and published with the support of the association's technical committees. This standard is currently being transferred to the international arena as part of an ISO standards project.

The Executive Board's strategy to extend its employee training programme to the European and international markets has been successful. The course contents, which were developed by the Fraunhofer IFAM for the various levels of qualifications (adhesive bonder, adhesive specialist, adhesive engineer) with the financial support of the association, have been translated into English, Chinese and other languages and adapted to the different applicable European and international standards. Training courses for adhesive bonders are now held on a regular basis in Poland, the Czech Republic, Turkey, the United States, China and South Africa. More than 11,250 adhesive bonders, specialists and engineers throughout the world have now been trained by the Fraunhofer IFAM team based in Bremen.

By developing this global training standard and putting in place a suitable worldwide training programme, the Executive Board of the German Adhesives Association has once again clearly demonstrated the key position of Germany's adhesives industry on international markets.

For IVK's Executive Board, providing sound training and qualifications in adhesive bonding technology is just as important as ensuring that young people have a good education in subjects such as chemistry, engineering and material sciences. Against this background, the contents and educational approach of the programme "Die Kunst des Klebens" (The Art of Bonding) have been comprehensively revised by experts from IVK, the German Chemical Industry Fund (FCI) and educational specialists. The course material is available to almost 20,000 technology teachers all over Germany.

Against the background of the initiative to digitise education in schools which has been adopted and launched by the German federal government, the Executive Board has decided to digitise its educational programme "Die Kunst des Klebens" (The Art of Bonding) and to link it to other information material provided by the association (the e-paper "Job Profiles", the promotional video, the adhesives guidelines, the adhesives magazine etc.). The association's partner for this project is the educational publisher Hagemann Bildungsmedienverlag, which used its technical expertise to digitise the educational material. As a result, teachers will have a tool that meets the new requirements for digital vocational education.

The adhesives industry is one of the first sectors of the chemical industry to offer teachers a digital package of this kind. This increases the chance of the subject of bonding and adhesives being included in the school curriculum. The digital teaching material was presented for the first time at the didacta 2018 fair in Hanover and received a positive response from the education profession. It was officially made available to schools at the start of the school year 2018/2019. The digital package was downloaded by many teachers during the first phase of the COVID-19 pandemic in particular, when all the schools were closed and teaching moved online. It was also used by parents for home schooling. The materials are still available to download free of charge: www.klebstoffe.com/informationen/unterrichtsmaterialien/.

In addition to this teaching material, the German Adhesives Association, in cooperation with FWU, the Institute for Film and Picture in Science and Education, has designed and produced two educational DVDs. The DVD entitled "Grundlagen des Klebens" (Adhesive Fundamentals) was developed for classroom use in schools and vocational colleges, while "Kleben in Industrie und Handwerk" (Adhesives in Industry and Trades) describes specific practical applications of adhesives involving different combinations of materials and is ideal for use in engineering and materials science teaching at vocational colleges. Both DVDs consist of a variety of films, animated sequences, interactive assessment tests and extensive information for instructors and students. The DVDs are available online from the FWU media library at www.fwu.de.

The subject of adhesives and bonding in teaching is covered in greater depth during the regular seminars for teachers held by the German Chemical Industry Association (VCI).

IVK is also increasingly becoming a leading player both in Europe and worldwide with regard to technical issues.

This applies in particular to the concept, jointly developed by the Executive Board and the Technical Board, of a standardisation competence platform in the German Adhesives Association. The objective of this project is to make use of the successful, in-depth involvement of the German Adhesives Association in European standardisation (CEN) and to play an active role in international standardisation activities at the ISO level. The main factors driving this initiative were, on the one hand, the increasing number of ISO standards for adhesives which are gaining growing acceptance as part of the globalisation of markets. On the other hand, the Executive Board is pursuing the long-term goal of ensuring that German industrial standards which are important for the adhesives industry become established globally. With its standardisation competence platform, the association has succeeded in setting up and implementing important standards projects in the fields of adhesives for floor coverings, wood and load-bearing bonded joints on an international level. In addition, it has produced key information for member companies about future developments in the electronics industry and the requirements for the relevant adhesives.

The German Adhesives Association has taken responsibility for managing the European standardisation project "Mandated flooring adhesive standard" and the European Standards Secretariat "Wood and Wood Materials".

The secretariat will protect the specific interests of the German adhesives industry in the complex regulated market for glued laminated timber for load-bearing applications.

The European Commission's Green Deal involves a wide range of ambitious measures to protect the climate, all of which are aimed at making Europe the first climate-neutral continent by 2050. The Circular Economy Action Plan and one of its key instruments, the Ecodesign Directive, are currently having an impact on the adhesives industry. The Ecodesign Directive creates a framework for establishing ecodesign requirements for energy-related products. Alongside energy efficiency, the important considerations for the European Commission include resource and material efficiency, durability, repairability and reusability. The Renovation Wave, the EU Industrial Strategy and the Chemicals Strategy for Sustainability are among other important measures that form part of the Green Deal and present both challenges and opportunities for the adhesives industry. In this context, adhesive bonding is often wrongly and unjustifiably subjected to criticism in relation to recycling and the circular economy. The Executive Board regards this as a clear signal that the industry would be very well advised to significantly expand its strategy for communicating with the EU and also with the technical departments of German ministries, other public bodies and associated institutions.

The responsible use of resources is also a fundamental requirement of the EU Construction Products Regulation, which concerns buildings and therefore construction adhesives. The construction and building industry generates a significant proportion of our greenhouse gas emissions and uses large amounts of energy. One prerequisite for sustainable development is an accurate life cycle assessment covering the environmental impacts of the construction products being used. Environmental product declarations (EPDs) are a standardised method of describing the life cycle assessment of a construction product. Adhesives with a variety of formulations are used for a wide range of applications in buildings and this makes the creation of individual, product-specific EPDs prohibitive for cost reasons. Therefore, a system of sample EPDs has been developed which cover specific formulations and applications and which allow the effects of the products with the greatest environmental impact to be specified. The sample EPDs can be used by all IVK members: https://www.klebstoffe.com/epd-nachhaltigkeit/.

Because of the importance of the Green Deal and sustainable building, the advisory board for sustainability was established on the initiative of the Executive Board and will take responsibility for handling these complex issues on behalf of the German adhesives industry. In addition, a sustainability communications strategy has been drawn up in close consultation with the advisory board for public relations. The aim of the strategy is to highlight the benefits and the potential of adhesive bonding as part of the solution for sustainability.

The fact that the German Adhesives Association and its members actively participate in important global conferences underlines the outstanding international position of the German adhesives industry. This applies in particular to the World Adhesives Conference (WAC) that takes place every four years. Following the international conference in Paris in 2012, member companies of the German Adhesives Association supported the 2016 World Adhesives Conference in Tokyo with a number of interesting presentations. The next WAC, which was due to take place in April 2020 in Chicago, has been postponed for two years because of the pandemic. The association has already prepared a block of lectures on the subject of sustainability and the circular economy for the conference that will now take place in the spring of 2022. Germany's adhesives industry and its association are therefore actively taking the opportunity to highlight and to document for an international audience the technological leadership and the skills profile of the industry in the field of sustainability.

During the Executive Board's regular interactions with adhesives associations in the United States and Asia, it is becoming increasingly evident that the German adhesives industry is not only considered to be a global technological leader but is also regarded with respect for its competence in "Responsible Care®" and sustainable development. In consultation with its system partners, the German adhesives industry began several years ago to develop practical solutions to provide adequate protection for the environment and consumers and to ensure workplace safety, before the relevant legislation was introduced, and went on to successfully implement these solutions in line with the strategic guidelines issued by the Executive Board. By separating solvent consumption from the production of adhesives, successfully establishing the EMICODE® and GISCODE systems and introducing sample EPDs, the German adhesives industry has fulfilled its responsibility to the environment and its customers throughout the entire value chain and is therefore playing a leading role on a global scale.

With voluntary initiatives to remove solvents from parquet adhesives and phthalates from paper/packaging adhesives and with its information series "Klebstoffe im lebensmittelnahen Bereich" (Adhesives in the food industry), the German Adhesives Association has repeatedly set new standards for health and safety at work and environmental and consumer protection.

The Executive Board of IVK believes that maintaining a balance in the relationship between technological leadership and social competence is the key to the successful and credible positioning of the industry. This ensures that the association always has access to reliable and important information, for example relating to the future direction of draft legislation, and that members of the association will have the opportunity in the future to open up new markets in Europe and worldwide with products that comply with the relevant requirements on workplace safety and environmental and consumer protection.

The members of the Executive Board view the constantly growing number of participants at the various events organised by the German Adhesives Association and the many new member companies which have joined the association during recent years as a clear indication that the association is well positioned, provides a valuable and very practical service for the benefit of its members and is therefore highly attractive to the adhesives industry as a whole.

Technical Board (TA)

In its regular meetings in 2020 and 2021, the Technical Board focused on the specialist work of the technical committees, the subcommittees and the ad-hoc committees, discussed this in detail and developed scenarios for the entire adhesives industry. Furthermore, the Technical Board held in-depth discussions on a number of interdisciplinary topics and issues which were then proactively influenced by the association's corresponding activities.

A major priority of the Technical Board's work has been the subject of the **EU Commission's Green Deal. The Green Deal** consists of 50 strategic and legislative measures that are intended to ensure that the EU is climate-neutral by 2050. The measures that will have the greatest impact on the adhesives industry are the New Circular Economy Action Plan (Transition to a Circular Economy), the Chemicals Strategy for Sustainability and the Clean Air and Water Action Plans (A zero pollution Europe).

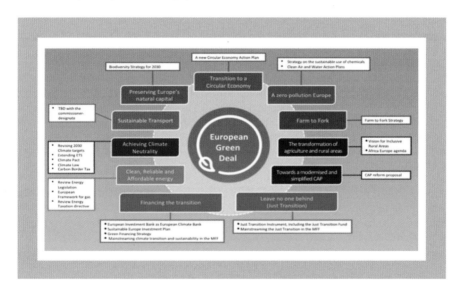

The EU's Circular Economy Action Plan (CEAP) is a package of linked initiatives aimed at creating a robust, coherent framework for product policy that will make sustainable products, services and business models the norm and change patterns of consumption to ensure that no waste is produced.

The focus of the new CEAP is on the following product groups: electronics, information and communication technology (ICT), batteries and vehicles, packaging, plastics, construction and buildings, food, water and nutrients and also textiles. It will also take into consideration furniture and interim products with a high environmental impact, such as steel, cement and chemicals.

The CEAP also lays the foundations for broadening the scope of the Ecodesign Directive to cover as wide a range of products as possible with regard to:
- Improving durability, reusability, upgradability, repairability and the handling of hazardous chemicals, plus increasing energy and resource efficiency
- Increasing the recycled content of products while guaranteeing their performance and safety
- Enabling remanufacturing and high-value recycling
- Reducing the carbon and environmental footprint
- Restricting the use of disposable products and introducing measures to prevent premature obsolescence
- Banning the destruction of unsold, non-perishable goods
- Introducing "products as a service" or similar models where manufacturers remain the owners of the products or take responsibility for their performance during the entire life cycle
- Digitising product information, for example digital product passports, marking and watermarks
- Recognising products on the basis of their sustainability, including by creating incentives

In order to identify the role of adhesive bonding in the context of the circular economy and life cycle assessments, the German Adhesives Association commissioned a study by the Fraunhofer Institute for Manufacturing Technology and Advanced Materials IFAM. The sustainable use of resources in the manufacturing, use and disposal of a product is determined by political policy and is also required by society as a whole. Specific targets have been set for the use of resources, the production of waste and emissions and the efficient consumption of energy. The tools for achieving these targets include long-lasting designs, maintenance, renovation, repairability, reuse, remanufacturing and recycling. Therefore, materials and joining technologies need to be developed within the industry to reduce resource use and avoid a linear economy. Adhesive bonding, which is a low-heat joining method that does not damage materials, can play a key role in this respect. The Technical Board contributed to the content of the study "Circular Economy and Adhesive Bonding Technology" which was published in August 2020.

Another project organised by the Technical Board involves the development of **sample EPDs** (environmental product declarations) for the building industry. This work was necessary because since 2011 the EU Construction Products Regulation has specified that "sustainable" products must be used and that the sustainability of products must be substantiated by EPDs. The EPDs, which are required by law, include detailed analyses and documentation, for example with regard to CO_2 emissions during production, energy consumption, resource depletion, life cycle assessments etc. The preparation of EPDs involves considerable work and substantial costs, which means that an industry-wide approach is the ideal solution. The German EPDs have been completed and offered to other European countries by FEICA.

EPDs have to be revised every five years, which means that the German versions need to be updated. The FEICA EPDs will expire between September 2021 (polyurethane adhesives) and August 2022 (dispersion adhesives). If the German EPDs and the FEICA EPDs are amalgamated, Germany will benefit from the longer lifetime of the FEICA versions. In addition, only one system will have to be revised and the costs can be shared.

An analysis of the revised version of EN 15804 "Sustainability of construction works. Environmental product declarations. Core rules for the product category of construction products" produced the following results:

- As the FEICA model EPDs are based on weight units, they are not affected by the change in the rules for function units.
- The FEICA model EPDs meet the requirements of the special rule, which means that the use phase (module C) and the end-of-life phase (module D) do not need to be taken into consideration. Transport on site and installation (A4 and A5) are already included in the existing EPDs.
- Environmental impact indicators: The FEICA model EPDs are based on the following environmental impact indicators: GWP (global warming potential), ADPF (abiotic depletion potential fossil fuels) and POCP (photochemical ozone creation potential). These form the basis for the evaluation of the individual substances. This pragmatic approach must be defended if necessary, particularly as the revision of EN 15804 will lead to the introduction of toxic criteria.

Another focal point of the Technical Board's work over the past two years included the activities relating to the European chemicals legislation REACH (Registration, Evaluation, Authorisation and Restriction of Chemicals), set out in the REACH Regulation (EU) No 1907/2006.

On 11 January 2019, the European Chemicals Agency (ECHA) published a restriction proposal for microplastics under the terms of REACH. The focus of the proposal is on intentionally added microplastics. A public consultation on this subject was launched on 10 March 2019 which lasted for six months.

The restriction proposal referred to above applies to microplastic concentrations ≥ 0.01% (weight by weight) defined as "solid polymer particles 1 nm ≤ x ≤ 5 mm" and fibres with a length of 3 nm ≤ x ≤ 15 mm and a length-to-diameter ratio of > 3, where the term "solid" is based on the definition in Annex I of the CLP Regulation and is almost entirely unsuitable for polymers. Products of all kinds are affected. The ECHA definition of "microplastics" includes among other things:

- Dispersions
- Fibre-reinforced and hardened cement-based products
- Spacers
- Granulated hotmelts
- Microballoons (adhesive tapes)
- 3D printing powder
- …

The draft restriction proposal is currently being revised. The existing lower limit of 1 nm is expected to be raised to 100 nm. Test methods will be added to the definition of "solid" and polymers which are (partially) soluble in water will be excluded (derogation from the scope). In addition, new rules will be introduced for artificial turf (sports pitches), foods, sewage sludge and compost (derogation from the restriction). Individual transition periods will be extended after they have come into force. It will also be possible to provide labelling/user information in the form of pictograms on the packaging.

Regulation (EU) 2020/1149 amending Annex XVII to REACH, which involves the addition of number 74 **"Diisocyanates"** (products with a diisocyanate concentration of > 0.1%) to Annex XVII, was published in the Official Journal of the European Union on 4 August 2020. The regulation relates to the industrial and professional use of diisocyanates and products containing them, includes no derogations and requires training to be taken (up to three levels depending on the potential exposure, with the course repeated every five years) with proof in the form of certificate. The mandatory requirement for training must be indicated on the label from 24 February 2022 at the latest. The industry will provide the training materials and the courses will be run by experts in occupational health and safety and occupational medicine and will include e-learning components. However, there is no cross-border recognition of the training because the restriction only requires a minimum standard (details will be regulated on a national level). The transition period is 18 months for suppliers and 36 months for users after the publication in the Official Journal of the European Union on 24 August 2020. In other words, from 24 August 2023, there is an obligation to successfully complete a training course before using the products. A working group consisting of eleven members of the PU Exchange Panel, including FEICA, was set up in April 2018 under the leadership of ISOPA/ALIPA to draw up the training materials. An external consultant will be responsible for coordinating and creating the training materials in the form of e-learning and classroom courses. Specific modules concerning adhesives will be prepared with the help of FEICA. The materials will be translated into all EU languages in accordance with the requirements of the regulation. The aim is to complete the English and German versions of the training materials by October 2021 and then to translate them into other languages as soon s possible after that. ISOPA/ALIPA intends to charge for the training courses, while FEICA is aiming to provide free training for customers.

In addition, the European Commission issued a mandate to the ECHA on 28 January 2019 to identify and introduce a harmonised occupational exposure limit (OEL) for diisocyanates (for the schedule, see below). No OEL for diisocyanates has been proposed in the ECHA's Scientific Reports for Evaluation of Limit Values at the Workplace. Instead it has been recommended that the Committee for Risk Assessment (RAC) should draw up an exposure response. A significant reduction in the range of 0.1 to 1 ppb is expected.

The consequence of this would be the need for more measurements to be made in order to comply with lower detection limits and, if necessary, investments in additional extraction systems.

Another REACH-related issue concerns the **registration of polymers.** Currently REACH requires monomers but not polymers to be registered. However, under the terms of Art. 138(2) of the REACH regulation, the European Commission can in specific circumstances present legislative proposals for the registration of certain polymer types. The next REACH review is planned for 1 June 2022. In contrast to previous activities, it is now certain that the European Commission will submit a proposal to register polymers under REACH. At the end of 2018, the commission tasked the consultancy companies Wood and Peter Fisk Associates (PFA) with carrying out another study, which was published in June 2020. This led to a discussion among the members of the European CARACAL group (Competent Authorities for REACH and CLP) and a subsequent impact assessment by the commission. The information requirements for registration will be based on the regulation for substances.

The study carried out by Wood/PFA proposes in particular the establishment of criteria for polymers that require registration and the formation of polymer groups. If one of the criteria is met, the polymer must be registered. The study includes:

- Criteria for the identification of polymers requiring registration
- Proposals for the formation of groups
- Appropriate information requirements for registration
- A cost-benefit analysis of the registration requirements that could be used by the commission in a subsequent impact assessment

The critical points are the criteria for identifying polymers that require registration (PRR criteria), the definition of the substance identity, the formation of groups and the necessary polymer analysis and test requirements. In its own working group, FEICA has drawn up a table which assigns the polymers that are used in adhesives to the individual Wood/PFA criteria and demonstrates that many of these polymers would be subject to registration.

The objectives of the next strategic measures to be taken by the adhesives industry are:

- The correction of the PRR criteria
- The reduction of the registration requirements

The proposal submitted by IVK, which is backed up by examples, to exclude from the registration requirements polymeric precursors (in other words prepolymers) that react in industrial use to form polymers is now the subject of wide-ranging discussion.

One specific measure that is part of the European Commission's action plan for the Chemicals Strategy for Sustainability concerns the introduction of **mixture assessment factors (MAFs).** Under the terms of REACH, all substances in mixtures can currently be "safely used" in accordance with the exposure estimations in the exposure scenarios. "Safe use" means that the RCR (risk characterisation ratio) is less than 1. The RCR is always below 1 if the expected exposure is less than the concentration at which no adverse effects on humans and the environment are likely (the effect threshold on humans and the environment).

The European Commission and several member states are concerned that it is possible for individual substances to influence one another and increase their effect. The commission refers in particular to adverse effects caused by unintended mixtures of substances. It has therefore been proposed that REACH should be amended to introduce a MAF which the RCR will be multiplied by when evaluating the risk of an individual substance.

The introduction of a general MAF would result in mixtures being categorised as much riskier than had been the case in the past, because the RCR of each substance in the mixture would be multiplied by the MAF. For the mixture to be considered safe, all the RCRs in the mixture would have to be less than 1.

If the calculated RCR of a substance in a mixture is higher than 1 after being multiplied by a MAF, in the REACH registration dossier the manufacturer of the substance must either use more complex and detailed calculation methods, restrict the permitted concentration in the mixture, reduce the period during which the mixture can be handled or introduce additional risk management measures to bring the RCR back below 1. If none of these measures are successful, the manufacturer must advise against using the mixture.

Initial analyses carried out by CEPE (European Council of the Paint, Printing Ink, and Artist's Colours Industry) indicate that many adhesive raw materials would no longer pass the REACH safety evaluation. Ultimately, it would then no longer be possible to use them in adhesives. This means that adhesive manufacturers with their formulations would be particularly hard hit by restrictions on the use of adhesive raw materials (on the part of the raw material manufacturers) or by their possible discontinuation.

The first discussions on MAFs have taken place in the European CARACAL group. However, there are currently no concepts or specific statements available. Therefore, the European Commission has commissioned a study to shed light on the issues and make proposals. The study will be carried out by the consultancy company Wood.

The **Regulation on harmonised information relating to emergency health response** was published as Annex VIII of the CLP Regulation in the Official Journal of the European Union on 23 March 2017.

Annex VIII of the CLP Regulation introduces a harmonised obligation for notifying European poison centres about mixtures. The unique formula identifier (UFI) will be used to report almost complete formulations in very narrow concentration bands. Under the original regulation, the introduction applied (a) to mixtures for consumer use from 1 January 2020, (b) to mixtures for professional use from 1 January 2021 and (c) to mixtures for industrial use from 1 January 2022. As the regulation was still unworkable when the first reporting deadline was reached (January 2020), following the intervention of the industry the first reporting deadline for mixtures for consumer use was postponed by one year to 1 January 2021 in the first amending regulation.

The second amending regulation was published on 13 November 2020 in the Official Journal of the European Union and covered the cases highlighted by the industry where the text of the regulation could not be implemented as required. It also introduced interchangeable component groups (ICGs), for example, with regard to adhesives that contain the same mixture of raw materials from different suppliers. Certain interchangeable components can be grouped together in an ICG if specific requirements are met, such as identical categorisation and labelling. This measure is intended to reduce the number of new notifications and new UFIs. In addition, the first notification deadline for mixtures for consumer and professional use (1 January 2021) was confirmed.

On the one hand, CLP Annex VIII specifies harmonised information requirements and types of information submission and replaces non-harmonised national notification requirements for mixtures. On the other hand, there are limits to the harmonisation. For example, national product registers remain in place and information must still be submitted to them. There is no standardisation of the notification languages (for example notification in all the official national languages in Finland). Some member states currently charge for receiving notifications and these charges will remain unchanged. The authorities in the member states can request additional information after receiving a notification.

On its website at https://poisoncentres.echa.europa.eu/de/echa-submission-portal, the ECHA has published a table showing information about the notification language, the fees, submission via the ECHA portal or the member state's portal, the readiness of the member state to accept notifications via the ECHA submission portal and the period after notification when the product can be placed on the market. In addition, in March 2021, the ECHA published version 4.0 of its guidelines which includes all the changes from the second revision of Annex VIII. The adhesives industry is particularly affected by the new notification requirements because of the large number of formulations and mixtures that it uses.

There has been and still is a need for action with regard to the **European Biocidal Products Regulation** (EU Regulation No 528/2012). As part of its evaluation of active biocidal substances that are already in use, including in-can preservatives, the EU Commission is issuing approvals in the form of implementing regulations. In particular in the case of substances with sensitising properties, the annex of the implementing regulations for product type 6 (in-can preservatives) generally contains a clause that makes mandatory the additional requirements for labelling specified in Art. 58(3) of the Biocidal Products Regulation (528/2012). The decisive factor in determining the deadline for implementing the labelling requirements is the date of approval of the substance. Only then do the special provisions of the implementing regulation come into force. IVK has informed its members in a number of newsletters about the approval of various in-can preservatives, the resulting changes in labelling and the accompanying deadlines. As an additional service, it has made available a table on the IVK intranet which gives an up-to-date overview of the in-can preservatives that have been newly approved by the EU Commission. The table indicates whether the implementing regulation for the active substance used as an in-can preservative in adhesives contains new labelling requirements in accordance with Art. 58(3) of the Biocidal Products Regulation (528/2012) and when these must be implemented.

The Technical Board has also discussed the subject of **residues in adhesive packaging and their impact on the recyclability of the packaging.** While reuse or mechanical recycling are common practice for large industrial-sized packages, recycling smaller packs is much more challenging. Although the adhesives industry makes every effort to use recyclable packaging wherever this is technically possible, in some cases the use of packaging which cannot currently be categorised as mechanically recyclable is unavoidable. In order to provide customers and other stakeholders with transparent, standardised information on this subject, the Technical Board has drawn up a document entitled "Gemeinsame Position der Klebstoffindustrie zum Einfluss von Restmengenanhaftungen auf die Recyclingfähigkeit von Verpackungen" (The joint position of the adhesives industry on the influence of adhesive residues on the recyclability of packaging).

On the basis of an initiative by the German Federal Ministry for the Environment (BMU) to reduce VOCs due to the risk of summer smog, VCI was asked, together with other industrial associations, to explore the possibilities for reducing the use of VOCs. Driven by concern for the environment, this initiative has led adhesive manufacturers to reduce their solvent consumption significantly over the past few years, as an evaluation of the **IVK solvent statistics** shows. The Technical Board has addressed this topic in depth and was able to demonstrate with the aid of its solvent consumption survey that the target of reducing VOCs by 70 % by 2007 (based on the consumption figures in 1988) was in fact reached much earlier. The solvent statistics are compiled every two years and used as the basis for discussions with German and European agencies and institutions, for example when introducing new legislative proposals. The statistics are a very important instrument for communicating with agencies and authorities and are indispensable for documenting the environmental awareness of Germany's adhesives industry. The last survey of solvent consumption for 2019 showed that solvent use fell by 3 percent, while at the same time adhesive production increased by 12 percent.

Other topics discussed by the Technical Board included:
• Standards activities
• Health and safety at work, environmental and consumer protection issues
• Monitoring various EU Commission projects in Germany

Technical Committee
Building Adhesives (TKB)

Introduction

The Technical Committee Building Adhesives (TKB) of the German Adhesives Association (IVK) represents the interests of manufacturers of building adhesives and dry mortar systems who are members of IVK. The committee liaises with public authorities, trade bodies, employers' liability insurance associations, other industry organisations and standardisation committees.

Its aims are to establish technical standards, to influence the provisions of legislation on chemicals, to participate in developing legislation, to promote technical progress while safeguarding users and the environment, to provide technical support and information to customers and the building trade, and to promote the use of building adhesives and mortar systems by providing objective technical information.

Overview of topics

TKB's activities can be broken down into several categories:
• Technical topics relating to building adhesives/flooring installation products and their applications.
• Standards for flooring and parquet adhesives, levelling compounds, tile adhesives, binders for floor screeds, primers and special products.
• Technical information events for the flooring and parquet laying trade and other related trades.
• TKB publications about current topics relating to application methods and building legislation and about standardisation and environmental and user protection issues.

- Building legislation issues, including approvals in Germany and Europe.
- Topics relating to chemicals legislation, such as German, European and international regulations on labelling hazardous materials and REACH.
- Health and safety at work and environmental and consumer protection.

The committee's work

Mortar systems
On the national and international standards bodies ISO/TC 189/WG3, CEN/TC 67/WG 3, CEN TC 193/WG 4, CEN TC 303/WG 2, NA 062-10-01 AA and NA 005-09-75 AA, representatives of TKB help to define the standards for flooring/parquet adhesives, tile adhesives, levelling compounds, sealing compounds and floor screeds.

The requirements standard (EN 12004-1) for tile adhesives was published in April 2017. However, it has not yet been issued in the Official Journal of the European Union, which means that Annex ZA of the old standard remains valid for CE marking. A new draft standard has now been produced which complies in full with the legal requirements of the European Commission, even though this reduces its technical relevance. Other important technical properties will be covered in part 3 of the standards series, which will not come under the umbrella of harmonised standardisation. Whether this draft will ultimately be accepted by the European Commission remains to be seen.

The situation for sealing compounds is similar. The draft of EN 14891 was revised in accordance with the requirements of the European Commission and part 2 of the standard will be drawn up with non-harmonised properties. In this case too, the European Commission has not yet responded. The draft versions of the revised test principles for sealing compounds have been published on the website of the German Institute for Building Technology (DIBt). However, because of the formal processes required, these will not come into force until at least 2022.

Product and application technology
TKB continued its comprehensive investigations into the subject of screed drying, the readiness of screeds for the installation of floor coverings and subfloor moisture measurement methods. The data presented at the specialist conference of the German Federal Association of Screeds and Floor Coverings (BEB) in 2019 concerning the comparison between CM and KRL measurements has now been published as IBF report M106/18 on the website of the IBF (Institute for Testing Construction Materials and Research into Flooring). The high-quality measurement and test data led to excellent new findings, which were either not acknowledged or were misinterpreted by the authors. It was possible to demonstrate that in the case of cement screeds there is no correlation between the dryness and the CM figure and that the results of the CM measurement cannot be assessed independently of the cement content. For this reason, parts of the results were re-evaluated by TKB and the findings of this process were published as TKB report 7 "An additional evaluation of the measurement data in the IBF report 'Investigations into the suitability of the KRL method for determining the moisture content of screeds'". As a result of this, the IBF issued an explanatory technical information sheet to accompany report M106/18. TKB responded

to this with TKB report 7a "Comments on the IBF technical information sheet 01/2021 dated 22 March 2021". In the meanwhile, the drying behaviour of cement screeds has also been investigated by means of parallel measurements of moisture levels using the CM and KRL methods. The results showed that systematic measurement errors with the KRL method caused by the use of the KRL cup could be almost entirely prevented. They also demonstrated that the relative standard deviation of the CM measurement is at least three times higher than that of the KRL measurement. The results were summarised and published in TKB report 8. A KRL cup that can be 3D printed was designed to allow the KRL measurements to be carried out correctly on building sites. TKB members obtained 200 of these cups, which will primarily be used by their application technology departments. Selected experts were supplied with KRL measurement cups free of charge.

In collaboration with the Heimtex Association, further fire tests were carried out at the TFI (Institute for Research, Testing and Certification of Floor Coverings and Textiles) on floor coverings with five different adhesives. No clear correlation could be identified between the strength of the adhesive and the fire behaviour. Heat resistance tests will now be carried out with these adhesives in order to determine a potential correlation with this property.

Standards

In cooperation with the IVK standards competence centre, TKB has helped to define international, European and German standards as part of the ISO/TC 61/SC 11/WG 5, CEN/TC 193/WG 4, NA 062-04-54 AA and NA 005-09-75 AA standards bodies.

The work of TKB on bringing the content of EN 14259 (Adhesives for floor coverings) and EN 14293 (Parquet adhesives) up to the ISO level was completed with the publication of the revised DIN EN ISO 17178 standard and the new DIN EN ISO 22636 standard. The ISO standards project concerning the determination of the emission behaviour of adhesives on the basis of the GEV testing method is currently underway. A draft ISO standard for toothed scrapers has been produced which is in the final agreement phase.

At the CEN, a standards project concerning the determination of the moisture level in subfloors has been launched. It describes three different processes, including the KRL method, all of which use the measurement of relative ambient humidity to indicate the moisture content of the subfloor. The resulting draft standard E-EN 17668 was published in June 2021.

In the case of levelling compounds, TKB representatives are involved via NA 005-09-75 AA and CEN/TC 303 in revising EN 13813. The German draft version of the revised standard was published in March 2018. The EN 13892-9 requirements standard relating to the shrinkage of screeds has been integrated into the EN 13813 standard. This standard has been approved by the CEN technical committees, but has not been published by the EU Commission in the Official Journal of the European Union. The reason for this is the same as in the case of EN 12004 and EN 14891. On the basis of measurements in accordance with EN 13892-9, shrinkage classes for mineral screeds have been integrated into the revised DIN 18560-1 standard. Cooperation between CEN/TC 193, which drew up the test standards for levelling compounds, and CEN/TC 303 is guaranteed by the membership of the technical bodies (liaison). In the DIN 18560 standards series, work on revising DIN 18560, Part 2 (Floor screeds on insulation layers) was completed and it will now be published as a draft standard. Good progress is being made on the new part 8 of this standards series (Decorative screeds). In the case of DIN 18560, Part 1, TKB has requested that the extension of the test for dimensional stability to 90 days be deleted, but a decision on this has not yet been made. Although levelling compounds can be tested in the same way as screed mortars and be given a CE mark, they are primarily used as levelling layers on screeds. In order to make a clear distinction between levelling compounds and screeds, TKB has launched a project to develop an application standard for levelling compounds via DIN NA 062-10-01 AA.

Events and publications

The 36th TKB conference, which was planned for March 2021, was postponed until 9 June 2021 because of the COVID-19 pandemic. As the situation in June still did not allow for face-to-face meetings, the conference was postponed again until 16 March 2022. The main theme of the event will be special structures. A panel discussion and a presentation of the legal evaluation of special structures are planned. Other themes of the conference include: machines and tools for removing old floor laying materials, screed additives – truth and reality, zoning – laying modern modular floor coverings, climate conditions on construction sites and their impact on floor laying materials, stability of floor laying materials for long-lasting floors – in-can preservatives in aqueous dispersion adhesives. Despite the two postponements of the TKB conference, the TKB has maintained the flow of information within the industry by developing a new digital event "TKB Update 2021" with two main themes, which took place on 9 June 2021. The first part of the event consisted of the report of the TKB chair on TKB's activities during the previous year and presentations of the TKB technical information sheets 18 (KRL method), 19 (Decorative screeds) and 20 (Special structures) by the respective working group leaders. The second part took the form of a panel discussion involving representatives of the TKB and two editors of industry magazines which covered current issues in the flooring industry. The participants felt that this part of the TKB Update 2021 was particularly interesting and informative. Despite the unfamiliar format, all the participants and the guests on the screen considered the event to have been a success.

The revision of the TKB technical information sheets has been completed. Technical information sheet 18 has been revised and now includes the requirements for suitable measuring devices, a list of suppliers and details of the KRL measuring cups. Technical information sheets 19 and 20 have been newly published. Like TKB technical information sheet 8, BEB information sheet 4.1 concerns the preparation of subfloors. Work on combining the two information sheets is ongoing. The Association of the Austrian Chemical Industry (FCIÖ) has adapted TKB technical information sheets 1 – 17 to the regulations in Austria. They will now be published as FCIÖ information sheets on the association's website. Information about the new TKB reports 7a and 8 has been provided above.

Statements by TKB on current topics, including the new TKB technical information sheets and reports are published under the heading "TKB informiert…" (Information from TKB) in the industry press and on the IVK website.

Cooperation with other associations/institutions

TKB consults with a variety of associations and organisations on technical and regulatory issues. An informal contact was maintained with the German Federal Association of Screeds and Floor Coverings (BEB) and with the Association of Manufacturers of Vinyl and Linoleum Floor Coverings (FEB) via some manufacturers of floor laying materials which are also supporting members of this organisation.

The collaboration with the German Association of Parquet and Flooring Installation Companies (BVPF), which is the most important body in this sector, proved to be very useful, largely due to the involvement of TKB in the BVPF expert advisory committee. The current important issues include the definition of the readiness of screeds for the installation of floor coverings and the measurement of the moisture content of screeds.

On a European level and in particular as part of the FEICA Working Group Construction, TKB representatives have contributed to the implementation of the Construction Products Regulation and supported the introduction of European standards for floor laying materials. The German sample EPDs that are now being used on a European scale are currently being revised and will be available with a new structure by the end of the year.

The project "Praxisgerechte Regelwerke im Fußbodenbau" (Practical regulations in flooring construction – PRiF) was started by the Federal Group for Screeds and Floor Coverings (Bundes-fachgruppe Estrich und Belag). The associations involved intend to recognise each other's information sheets with the aim of establishing generally agreed technical regulations. TKB is an active participant in the project.

Issues relating to building legislation

Although the majority of those involved proposed no changes or only minor changes to the Construction Products Regulation, the European Commission is planning a detailed revision of the legislation. The commission was critical in particular of the cooperation with CEN. A first draft of the revised Construction Products Regulation is expected in 2021. No definitive dates have been given for its entry into force, but 2025 has been mentioned.

Hazardous materials/health and safety at work

As silane-based primers can produce higher levels of methanol emissions than silane-based adhesives because of the different application conditions, BG Bau (the German employers' liability insurance organisation for the construction industry) is introducing a new GISCODE group "RS 20 – Floor laying materials, primers that contain methoxysilane".

The restriction on the use of products containing diisocyanates came into force in August 2020. As a result, the affected products must have a label on their packaging from February 2022 and users can only work with these products from August 2023 if they can prove that they have taken part in training. The training documents are being produced by the raw material manufac-turers. It is likely that both classroom and online training will be available.

The European Commission is working on the registration of polymers as part of REACH. There is very little definitive information about this, but TKB is included in the information flow via the VCI.

Technical Committee
DIY and Consumer Adhesives (TKHHB)

The Working Group and Technical Committee DIY and Consumer Adhesives (AGHHB/TKHHB) regularly hold joint meetings and monitor a variety of legislative activities on a European and national level. Their main interest is in regulations and topics concerning adhesives in small packages which are intended for private end users. Important topics in this area include:
- Identification and packaging of adhesives and the obligation to provide information (CLP Regulation)
- Requirements in certain areas where adhesives are used (Appliance and Product Safety Act/ Toy Safety Ordinance/Medical Products Act)
- Other standards-related/legislative restrictions and requirements regarding adhesives

Labelling and packaging of adhesives and the obligation to provide information
(CLP Regulation 1272/2008)
According to Article 4 of the CLP Regulation, the manufacturer or importer must:
- classify substances and preparations prior to marketing them
- package them according to the classification and
- label them.

The Technical Rules for Hazardous Substances (TRGS 200) summarise the relevant regulations:
- special labelling regulations for substances and preparations that are available to the general public, 6.7; 10.2
- simplification of labelling requirements and exceptions, 7.1
- implementing labelling regulations, 9

Additions to the CLP Regulation resulted in new obligations for manufacturers.

A new Annex VIII was added to the CLP Regulation in March 2017 which covers the implementation of requirements for harmonising information in notifications in accordance with Art. 45. This is mandatory in all member states from 1 January 2021. The key elements of the harmonisation process are the standardised notification format (PCN format) and the option of submitting notifications via a central European Chemicals Agency (ECHA) portal.

Annex VIII defines three different categories for the use of mixtures. There are different notification deadlines for each category of use which determine when the new rules must be applied:
1. For mixtures for consumer use: from 1 January 2021
2. For mixtures for commercial use: from 1 January 2021
3. For mixtures for industrial use: from 1 January 2024

The national regulations continue to apply until the notification deadlines. In Germany, transitional regulations are specified in section 28 paragraph 12 of the Chemicals Act (ChemG).

The main new feature of the harmonised notification is the unique formula identifier (UFI). This 16-character alphanumeric code will be assigned to every mixture that requires notification and must appear on the label or the internal packaging. The UFI can be specified on the safety data sheet (SDS) for unpackaged mixtures or mixtures for industrial use.

In Germany, notifications in the new PCN format can be sent to the Federal Institute for Risk Assessment (BfR) or submitted to the ECHA via a central portal. The ECHA sends the notifications submitted to it which are destined for Germany to the BfR. In this case, an effective notification is only considered to have been submitted when it has been received by the BfR. No charge is made for this process in Germany.

Companies can obtain comprehensive information about Annex VIII of the CLP Regulation from the following page on the ECHA website:
https://poisoncentres.echa.europa.eu/de/steps-for-industry.

Requirements in certain areas where adhesives are used (Medical Products Act/Appliance and Product Safety Act/Toy Safety Ordinance)
As a general rule, a product may only be sold if it is designed in such a way that the health and safety of users or third parties are not at risk if it is used as intended or in a manner that can be foreseen.

This affects the manufacturers of items that are ready for use. Adhesives are not directly subject to this regulation but are indirectly affected via the end product if the safety (and usability) of an item depends on the suitability of the adhesive.

Adhesives as toys or components in toys for the purpose of the Toy Safety Ordinance/ Appliance and Product Safety Act – Toy Safety Directive – EN 71.
EN 71 was revised on behalf of the EU Commission. It is based on the safety requirements of the Toys Directive 88/378/EEC, which requires toys, in other words products which are intended for children younger than 14, to be safe before they are placed on the market. Compliance with the requirements defined in EN 71 is documented by the CE mark. Tests can be carried out by the manufacturer or by a testing body. Alternatively an EU type approval test can be performed if compliance with EN 71 cannot be established in any other way.

Product information in emergencies (DIN EN 15178), labelling checklist
The purpose of the product information standard DIN EN 15178:2007-11 is to improve the identification of products in the event of emergencies. The letter "i" placed near the bar code on packaging refers to the trade or product names or the number under which the product is registered or officially approved.

The Technical Rules for Hazardous Substances (TRGS 200) "Classification and identification of substances, preparations and products" provides another template for a checklist for labelling issues, which has been revised by TKHHB.

New Packaging Act

The AGHHB has also been working on the Packaging Act, which was amended in 2021 and which regulates the marketing and supply of packaging and the recovery and high-quality recycling of packaging waste. Manufacturers, distributors and importers which are the initial suppliers in Germany of B2C packaging that is subject to the system participation obligation must join a (dual) system to ensure the nationwide recovery and recycling of the relevant packaging waste.

In a purely commercial context, there is no requirement to take part in the system. This makes it possible to use a more cost-effective commercial system for the recovery or to have the packaging recovered by the supplier.

A declaration of completeness, registration details and notification of the used packaging must be submitted to the Zentrale Stelle Verpackungsregister (Central Agency Packaging Register – ZSVR). In September 2021, the ZSVR plans to publish new details of the minimum standard for assessing the recyclability of packaging that is subject to the system participation obligation. The current regulations date from 2019. Financial incentives will be used to reward companies for using packaging with a particularly environmentally friendly design.

The second amendment to the Packaging Act (VerpackG) imposes many new obligations on companies.

- Completely empty packaging that previously contained hazardous substances must be disposed of separately, but only if the packaging is contaminated. The definition of "contained hazardous substances" is now based on the current version of the Banned Chemicals Ordinance (ChemVV) from 2017.
- Packaging that is not subject to the system participation obligation (commercial or industrial packaging) must now also be registered with the ZSVR.
- From January 2021, products that contain > 0.1% substances of very high concern (SVHC) must be reported to the SCIP database. This also applies to packaging.

The Working Group and Technical Committee DIY and Consumer Adhesives continuously monitor the amendments to the Packaging Act and keep the association's members regularly informed.

Construction legislation

New regulations in the field of construction legislation also affect the working group. National regulations on construction products and on labelling, like those in France and Belgium, and approval regulations issued by the German Institute for Building Technology (DIBt) in Germany are important, because in some cases they affect DIY products. A VOC classification system is being considered on a European level that will be incorporated into the declaration of performance (CE) of the products via harmonised standards.

Under the coordination of the DIBt, the German federal states have restructured the approval regulations for construction products with the introduction of a new Model Building Regulation and the development of the Model Administrative Provisions – Technical Building Rules (MVV TB). Annex 8 of the MVV TB applies health-related requirements to floor covering adhesives and structural adhesives, although the legal situation (Judgement dated 7 October 2020 by the

Mannheim Higher Administrative Court on OSB panels) and the labelling obligation (mark of conformity) remain unclear.

On an EU level, the harmonised standardisation process has come to a standstill. The European Commission aims to produce a revised version of the Construction Products Regulation (CPR) by the end of 2022 and to make more targeted use of the various sustainability programmes of the Green Deal.

Since 2018 construction contracts in Germany have no longer been covered by the legislation on work contracts, but have their own provisions (sections 650a ff. of the German Civil Code). As a result, clients have additional obligations and rights, for example the specification of a completion date and a 14-day cancellation period. In addition, construction companies must provide comprehensive information about plans, approvals and certifications, for example. If individual services do not correspond with what was promised, the construction company will be responsible, unless a clear statement to the contrary was made in the construction contract. Provisions have also been introduced concerning the amounts of payments by instalment.

Technical Committee
Wood/Furniture Adhesives (TKH)

One of the main activities of the Technical Committee on Wood Adhesives (TKH) involves performing a consultancy and monitoring function in many areas of the wood processing industry. Health and safety at work, environmental protection and consumer protection are key considerations in this respect and have increasingly taken on a central role in recent years. This includes the issue of preservatives, where new thresholds and regulations have come into effect which our member companies have had to implement. The close, positive contacts that the TKH has with associations and institutes in Germany and in other countries are very important and over the years have been of valuable help.

The TKH expert panel is currently working on a new series of data sheets on the subject of avoiding delamination in various applications and using different types of adhesive. Several meetings have already been held on the subject of bonding to solid wood, where the influence of factors such as the manufacturing process, quality control and preparatory measures were considered in detail.

The existing series of data sheets covering specific adhesive-related subjects is available to download in German and English from the IVK website.

In addition to its internal activities, TKH also takes part in working groups covering a number of different industries, such as the profile wrapping working committee and the initiative group on three-dimensional furniture fronts.

The members of the profile wrapping working committee are also active members of CEN/TC 249 WG 21, which is currently working on the EN 12608-2 standard "Unplasticized poly(vinyl chloride) (PVC-U) profiles for the fabrication of windows and doors". This enables our members to play an active role in the development of this standard.

The draft of the VDI (Association of German Engineers) guidelines "VDI RL 3462-3 Reducing emissions – Wood processing and woodworking; Processing and finishing wood materials" appeared as a white print in July 2020. All the changes proposed by TKH during the revision process have been included. The guidelines have not yet been published.

TKH is regularly represented on the CEN TC 193 SC1 Adhesives for Wood standards committee and its subcommittees. The DIN EN standard 14257:2019 "Adhesives. Wood adhesives. Determination of tensile strength of lap joints at elevated temperature (WATT '91)" has been revised by CEN TC SC1/WG 12 and was published in December 2019. The formal approval of the standard project "Classification of thermoplastic wood adhesives for non-structural applications for outdoor use" prEN 17619 is planned to take place between 8 July and 2 September 2021.

It is very important that standards and proposed standards are monitored with regard to their practical relevance, so that it is possible to respond accordingly if required.

I would like to take this opportunity to thank the member companies that give their employees the opportunity to work on behalf of the association and use their collective expert knowledge within the TKH for the benefit of users.

Technical Committee Adhesive Tapes (TKK)

The publicly funded research projects supported by the Technical Committee Adhesive Tapes (TKK) are making good progress, but the restrictions resulting from the COVID-19 pandemic have had an impact on their work. Both the current projects have been delayed and the project deadlines have been extended to compensate for this.

The driving force behind these projects remains unchanged. They are motivated by requests from industries that use adhesive tapes, such as the vehicle industry, and also by the general recognition that, in comparison with liquid adhesives, for example, there are very few research projects on the subject of adhesive tapes with publicly available results. Ensuring that a broader range of research is carried out is an essential requirement for increasing the use of adhesive tapes in more technically challenging applications.

One project is being run by the Fraunhofer Institute for Manufacturing Technology and Advanced Materials (IFAM) in Bremen. This is entitled "Possibilities and limits for regulating the reaction rate in accordance with the Arrhenius relationship in the rapid ageing of pressure-sensitive adhesives". The project has been underway since June 2019 and was planned to last for two years. It has now been extended until the end of September 2021. The project has investigated the mechanisms of thermo-oxidative ageing and hydrothermal ageing of acrylate pressure-sensitive adhesives (PSAs). This includes developing a method to determine the reliable use of accelerated ageing and to identify the temperature limits. The aim is to establish an ageing test that can be carried out at the maximum possible speed without changing the ageing behaviour and that can be used in real-life applications. This is based on the assumption that the ageing mechanism of acrylate PSAs is largely independent of the commercial product in question. The results achieved so far from the hydrothermal ageing process show that the solvent-based acrylate systems are highly resistant to ageing both in the standard ageing test that was selected (90 °C/240 h) and in the accelerated test with an increased temperature of 120 °C.

In the tests, none of the adhesive tapes produced clear negative results. The characterisation methods used (IR, DSC, rheology; pyrolysis-GC-MS etc.) also confirm the high temperature stability of the PSAs that were tested. Finally, none of the PSAs failed after ageing during the characterisation process.

Therefore, the investigations of thermal ageing were continued using dispersion-based adhesives with a focus on thermo-oxidative ageing. The storage periods have been completed and the results of the accompanying analytical tests are being submitted for final evaluation. Following the assessment of these results, a joint workshop involving the committee that is supporting the project and the TKK will be held to access the complex results and draw up a proposal for how to proceed with them.

The second project also concerns the requirements of vehicle manufacturers and, in particular, the public transport sector. It involves the fire protection characteristics of adhesive tapes, including flame retardance and smoke properties. This project has the title "Simplified methods for evaluating the fire behaviour of adhesive tapes and pressure-sensitive adhesive bonds". The project started on 1 December 2019 and will last for 24 months. The research is taking place at the IFAM in Bremen and at the Federal Institute for Materials Research and Testing (BAM) in Berlin. This project was particularly hard-hit by the COVID-19 restrictions and has therefore been extended by nine months until the end of August 2022.

The aims are to reduce the flammability (occurrence of fire) and increase the fire resistance of products bonded with PSAs in order to gain a better understanding of how a pressure-sensitive adhesive tape (adhesive and carrier) must be structured and how a component must be designed to achieve these aims.

The project will attempt to identify simplified measurement methods for carrying out preliminary tests of fire behaviour during the development of tapes which will enable companies, including SMEs, to meet the increasing fire prevention regulations. During the project, an investigation will be carried out into the influence of the fire behaviour of the adhesive, the adhesive tape with the carrier, the bonded materials and the design on the fire behaviour (flammability, spread of the fire and fire-resistance) of the whole component.

To achieve this, the fire behaviour of the adhesives will be varied with specific aims in mind. In addition, carriers and substrates with different fire behaviours and thermal conductivity levels and different design variants will be used.

At this point in the project, tests of the variations in the fire behaviour of the pressure-sensitive adhesives have been carried out and different test methods for the fire behaviour of the tapes and the bonded substrates have been evaluated.

A completely new area of activity has emerged as a result of legislative plans on the part of the EU. These concern a possible requirement for the registration of polymers in the context of REACH. Currently REACH requires monomers but not polymers to be registered. However, under the terms of Art. 138(2) of the REACH regulation, the European Commission can in specific circumstances present legislative proposals for the registration of certain polymer types. The next REACH review is planned for 1 June 2022. Because of recent developments (including the plastics strategy and microplastics), in contrast to previous activities, it now seems certain that the European Commission will introduce regulations concerning polymers and it has already begun evaluating the possible options.

This highly complex subject is, of course, extremely important for the entire polymer industry and not only for adhesive tape manufacturers, but in particular for manufacturers that synthesise polymers themselves for their products or those of their customers. TKK is closely monitoring the discussions that are taking place in industry associations, the VCI and FEICA and developing concepts that will help to determine what is done. The focus is currently on technical and scientific questions, but issues concerning the regulatory framework for the processes are gaining in importance. Because of the huge variety of polymers, questions about the categori-sation and the identical or similar nature of substances, for example, which are very easy to answer in the case of low molecular polymers, can quickly become highly complex. Currently, proposals are being drawn up to allocate polymers to the group which will have to be registered under the terms of REACH (PRRs = polymers requiring registration) and the group which will definitely not have to register (PLCs = polymers of low concern). These proposals are being incorporated into other committee activities on a European level. A second area of work based on the first one consists of proposals for designing the processes for registering PRRs. This area is closely linked with another problem that concerns the question of the "identical nature" or more accurately the "similarity" of polymers, which needs to be answered before the PRRs can be put into appropriate groups and registered together. TKK is the member of a FEICA working group dealing with this issue.

TKK is also involved in activities concerning standards issues that relate to measurement processes on a European and an international level.

Afera (the European Adhesive Tape Association) has worked with the PSTC (the Pressure Sensitive Tape Council) and JATMA (the Japanese Adhesive Tape Manufacturers Association) to harmonise the different regional methods for measuring adhesive strength, shear resistance and tensile strength. Afera was responsible for submitting the new harmonised methods to the ISO secretariat for use in countries throughout the world. The harmonisation process has now been completed on a European level (CEN) and an international level (ISO 29862, ISO 29863 and ISO 29864). After several attempts by CEN, these ISO methods have now been implemented as EN and DIN methods.

As has already been reported, new test methods are being developed by the Global Tape Forum (GTF) which will subsequently be published as GTF test methods. In the GTMC (Global Test Methods Committee), the ISO test methods have been taken over as GTF test methods. The newly developed GTF test methods are as follows: Shear adhesion failure, GTF 6001; Thickness, GTF6002; Width and length, GTF 6003; Loop tack, GTF 6007. The ISO test methods have been given the numbers GTF 6004 – 6006.

For the sake of completeness, it is worth mentioning that the GTF and GTMC consist of members of the adhesive tape associations of China, Europe, Japan, Taiwan and the USA. This means that they represent almost the entire global adhesive tape industry, because there are no other associations of adhesive tape manufacturers.

The new Test Methods Manual for Adhesive Tapes is almost complete. The production of the manual has been delayed because it has been completely revised. A number of old test methods were removed, the remaining old test methods were brought up to date and the current GTF test methods were included. A "soft version" was distributed to members online once the text had been completed. The decision was made to create new illustrations of all the test methods to give the manual a standardised, modern appearance. The illustrations have not yet been finished because of the problems caused by the pandemic.

The four test methods that were deleted from the manual, which are also DIN EN test methods, have now been removed from the list of DIN EN test methods following a procedure involving CEN. The new manual will only be available online and will be free of charge for Afera members.

With regard to new test methods, the PSTC has shelved the planned development of a dynamic shear test method, because it did not have the necessary support from the organisation's members. Afera then decided to introduce this test method, because a survey among its members indicated significant interest in the adhesive tape industry. Around 95 percent of them agreed that it was needed. In order to get started quickly, the Afera TC is currently developing a new method based on the FINAT FTM18 test method, with the agreement of FINAT. A proposal has been almost completed and will soon be distributed to the TC members as a basis for discussions at the forthcoming meeting in September.

Another new subject under consideration by the Afera TC concerns future test methods. There are plans to include dynamic mechanical analysis and other complex methods.

The technical committee is becoming increasingly involved with environmental issues and is entering into closer cooperation with FEICA in this area. This also means sharing information with IVK on these subjects. The corresponding working groups have been set up within the Afera TC.

As part of its activities, TKK is continuing to support Afera. TKK also contributes reports on conferences and events organised by the adhesive tape industry and on new products to the Afera News newsletter.

Technical Committee
Paper and Packaging Adhesives (TKPV)

Adhesives for products intended to come into contact with food

As in previous years, the Technical Committee Paper and Packaging Adhesives continued to focus on adhesives for food contact materials during the 2019/2020 period. It is still the case that adhesives used for producing materials and articles intended to come into contact with food are not specifically covered by EU regulations. Due to the fact that adhesives are part of food contact materials, they are, however, subject to assessment under the provisions of food legislation (Framework Regulation (EU) No 1935/2004). If adhesives used for products that come into contact with food are based on materials which are also used to produce plastics for products that come into contact with food, information from the corresponding EU regulations can be used for a risk assessment based on Article 3 of Regulation (EU) No 1935/2004. In this context, Regulation (EU) No 10/2011 came into force on January 2011 and replaced Directive 2002/72/EC of 6 August 2002. It consolidated all the material lists from the annexes and all the changes and amendments in one regulation. This has been added to and/or corrected approximately twice a year since it came into force.

For substances that are not listed there, the national European regulations, such as the recommendations of Germany's Federal Institute for Risk Assessment (BfR) or the Royal Decree RD 847/2011, still apply. Regulation (EU) No 10/2011 states that adhesives may also be made of substances that are not approved in the EU for the production of plastics.

Regulation (EC) No. 2023/2006 on good manufacturing practices for materials and articles specifies the principles of good manufacturing practices (GMP), as required by Article 3 of Regulation (EU) No 1935/2004 (EU Framework Regulation for Materials and Articles Intended to Come into Contact with Food). This regulation applies to all articles and materials listed in Annex 1, including adhesives. Strictly speaking, the GMP regulation does not apply to raw materials that are used in the corresponding adhesives. Nonetheless, the materials do have to fulfil specifications which enable adhesive manufacturers to work in accordance with GMP. Guidelines entitled "Good Manufacturing Practices" are available to help adhesives manufacturers to comply with the regulation.

The Union Guidance on Regulation (EU) No 10/2011 was published on 28 November 2013 by the services of the Directorate-General for Health and Consumers. It is intended to help with interpreting and implementing questions about declarations of conformity and other conformity activities and with providing information along the supply chain for food contact materials. In this EU guidance document, adhesives for the manufacture of food contact materials made of plastic or plastic composites are described as "non-plastic intermediate materials". Section 4.3.2 lists all the relevant information that a manufacturer of non-plastic intermediate materials should provide throughout the supply chain.

TKPV technical information sheets 1 to 4 have been revised on this basis. The information sheets already cover all aspects of the Union Guidance on Regulation (EU) No 10/2011 with regard to adhesives. Nevertheless, a formal revision taking into account the Union Guidance and all the new regulations and directives was unavoidable. German and English versions of each information sheet are available on the association's website.

The subject of mineral oils in foods is still occupying German legislators, who would like to introduce new regulations for food packaging made of recycled cardboard. The main source of these oils has been identified as the mineral-oil-based printing inks from newspaper printing.

The fourth draft of the "Mineral oil ordinance" produced by the German Federal Ministry of Food and Agriculture for recycled cardboard has been completed and is currently in the European notification procedure (TRIS).

The draft states that food contact materials made from paper, paperboard or cardboard can only be manufactured from recycled fibre and distributed if they have a functional barrier which ensures that no mineral oil aromatic hydrocarbons (MOAH) can be transferred from the food contact materials into the food. A transfer is considered not to have taken place up to a detection limit of 0.5 milligrams of all MOAH per kilogram of food or food simulant. The limits originally proposed for mineral oil aliphatic hydrocarbons were removed, together with the concentration limits in cardboard.

The food industry and all the industries that supply food packaging producers are still facing the problem of false-positive results from analyses. Beeswax, vegetable oils and colophonium produce similar MOSH/MOAH results to mineral oil and can be difficult to distinguish from it.

However, in Regulation No 10/2011/EC, materials that contain mineral oil saturated hydrocarbons (MOSH) and/or mineral oil aromatic hydrocarbons (MOAH) are listed and approved for the production of food contact articles made from plastic. Examples are fully hydrogenated hydrocarbon resins, which are also used in polyolefins for plastic packaging.

The TKPV technical information sheet 7 "Low molecular weight hydrocarbon compounds in paper and packaging adhesives" provides comprehensive information about the materials used in adhesives which could be detected as MOSH/MOAH in migration analyses but which have been adequately analysed for food packaging. It also provides guidance on selecting suitable raw materials.

Food Federation Germany publishes data at irregular intervals about the normal MOSH/MOAH content of different food groups (guidance values). This helps with the assessment of MOSH/MOAH findings and indicates whether the concentrations in the food are abnormally high.

Adhesives in paper recycling:
Another important and ongoing aspect of TKPV's work involves assessing the impact of adhesive applications on paper recycling. The current calls by the EU Commission and Germany's Federal Environment Ministry for higher recycling quotas for paper and the end of exports of "poor quality" waste paper to China will require new efforts to be made to improve the recycling process.

In this context, TKPV is taking part in a detailed dialogue with the paper industry and scientific institutions. With the support of TKPV, the International Research Group on De-Inking Technology (INGEDE) has developed the testing procedure "INGEDE Method 12 – Assessment of the Recyclability of Printed Paper Products – Testing of the Fragmentation Behaviour of Adhesive Applications", which has been incorporated into all the relevant eco-labels on a national and European level. The method evaluates the removability of an adhesive application from the recycling process using a screen. The method can be used for all applications of non-water-soluble and non-re-emulsifiable adhesives. Scientific studies which would allow all other types of adhesive application to be evaluated with regard to this recycling process have not yet been carried out.

In addition, TKPV has successfully completed a clustering project in cooperation with INGEDE. Adhesive applications no longer need to be tested using INGEDE method 12, if the adhesive has a minimum softening point and the application takes into account the minimum film thickness and the minimum horizontal expansion. The scorecard of the EPRC (European Paper Recycling Council), the guidelines for awarding Blue Angel certification RAL DE-UZ 195 and the EU Ecolabel have been amended accordingly. Afera (the European Adhesive Tape Association) has ensured that TKPV is represented on the EPRC.

The newly created Zentrale Stelle Verpackungsregister (Central Agency Packaging Register – ZSVR) publishes regularly revised versions of the minimum standard for recyclable design. This describes how packaging must be designed to achieve the highest possible recycling rate, but only takes into consideration existing recycling processes. The INGEDE 12 and PTS-RH:021/97 methods are used to assess adhesive applications in paper and cardboard packaging. CEPI, the European association representing the paper industry, has already published a harmonised European test method for evaluating the behaviour of paper products in the recycling process. In future this will replace all national methods. However, the accompanying evaluation method has yet to be developed.

REACH/CLP:
Another key area of TKPV's work involves activities relating to the new European Chemicals Regulation "REACH" (Registration, Evaluation, Authorisation and Restriction of Chemicals). Under the provisions of REACH, it is necessary to provide substance data and, in particular, exposure information to ensure that substances are handled safely. For this reason, downstream users of substances must inform the registrants about their use. In order to guarantee that this communication takes place, ECHA has developed a "Use Descriptor" model. On the basis of this model, TKPV has developed use scenarios for the production and the common applications of paper and packaging adhesives. The use scenarios have been included in the FEICA use maps for adhesive production and application.

The concentration limit for labelling as a hazardous substance under the current provisions of the CLP Regulation for methylisothiazolinone as an in-can preservative have been changed to correspond with the existing limits for chloromethylisothiazolinone/methylisothiazolinone (3:1). Because products that require no labelling now fall below the limit for methylisothiazolinone, many producers have moved to using benzisothiazolinone.

Standards activities:
TKPV has been successful in ensuring that German manufacturers of paper and packaging adhesives have expert representation on the Adhesives for Paper, Cardboard, Packaging and Disposable Sanitary Products work group of the European standardisation committee CEN/TC 193 and the German mirror committee DIN/NMP 458.

For ten years, the IVK Technical Board has been monitoring a number of DIN, CEN and ISO standards committees and mirror committees, including those referred to above. At the TKPV meeting on 13 June 2019 in Cologne, TKPV decided to bring an end to its own monitoring activities and to obtain information about any changes from the Technical Board.

Networking:
IVK has taken out a subscription with the Federal Institute for Risk Assessment (BfR) on behalf of TKPV to ensure that TKPV is informed directly about changes to the recommendations for food contact materials. In addition, at TKPV's request IVK is now a member of Food Federation Germany, which gives it rapid access to a wide range of information sources. TKPV has a permanent representative on the Food Federation Germany working group for food contact products.

FEICA:
Due to their importance at a European level, the topics of adhesives for materials and articles intended to come into contact with food and adhesive applications in the recycling process (paper and plastic) continue to be discussed extensively with the European Paper and Packaging and TF SRAPPA working groups at FEICA. The aim is to establish a joint position for the European adhesives industry and to find solutions to these important issues. TWG PP has also investigated mineral oils in foods and has produced technical information sheets on the subject. The working group has investigated a variety of raw materials used in hot melts for the packaging industry in order to determine the migration potential of mineral oil hydrocarbons. The results have already been presented in a webinar in 2021.

Other topics dealt with:
- The work of the Postpress Technical Committee of the Research Institute for Media Technologies (FOGRA)
- The research forum of the PTS
- The work of the Industry Association for Food Technology and Packaging (IVLV) in relation to food packaging
- Microplastics
- Biocides as in-can preservatives for aqueous adhesives
- The EU Green Deal

Technical Committee
Footwear and Leather Adhesives (TKS)

The Technical Committee Shoe Adhesives (TKS) coordinates public relations activities on behalf of German manufacturers of adhesives for footwear materials. In addition, the committee supports German and international standardisation activities and acts as the point of contact for technical evaluations and information specific to certain market segments as part of the work of the German Adhesives Association (IVK) and the German Chemical Industry Association (VCI) with regard to regulatory issues.

Standards activities
The committee focuses on supporting the development of German and European standards to define the basic properties of footwear adhesives.

These activities include standards for:
• Minimum requirements for footwear bonds (requirements and materials)
• Tests to check bond strength in footwear (peel resistance tests)
• Processing (determining the ideal activation conditions and sole positioning tack)
• Durability (colour change caused by migration, thermal resistance of lasting adhesives)

The provision and availability of standardised reference test materials and reference test adhesives are crucial for the ability to perform a large number of tests. The selection and specification of the reference test materials and adhesives are continuously evaluated and updated on the basis of the latest technology. The standard was revised in 2018/2019 and is now undergoing the approval process in the European committees. The TKS information sheet entitled "Problemlösung bei Fehlern in der Schuhherstellung" (Troubleshooting in footwear manufacture) is available to download from IVK's website. It provides assistance and tips for emergencies and service issues that occur during the use of adhesives.

Training
Providing training in adhesive bonding for employees in the footwear industry is another of the tasks that TKS is responsible for. The first courses took place in Zweibrücken in 1990 and in Pirmasens in 1992. TKS aims to share adhesives knowledge that makes it possible to identify practical problems more quickly, develop solutions and take measures to minimise and even eliminate potential sources of faults in the future.

Against this background, TKS has developed a training concept in conjunction with IFAM/Bremen and PFI/Pirmasens. It takes the form of a practical seminar called "Angewandte Klebtechnik in der Schuhindustrie" (Applied adhesive technology in the footwear industry). It was held for the first time in November 2005 with more than 20 participants. They gained an in-depth understanding of adhesive technology and the associated issues. The training focuses on providing practical examples to demonstrate best practices in bonding and typical problems. The participants can apply what they have learned in practical exercises, which makes the theoretical information easier to understand for users working in shoe production.

Technical Committee
Structural Adhesives and Sealants (TKSKD)

In the light of the growing number of adhesive applications in which adhesive bonds play a structural role, the Technical Committee Structural Adhesives and Sealants (TKSKD) has been focusing on a range of current technical issues relating to structural adhesives and sealants in order to support the manufacturers of structural adhesives and sealants which are members of the corresponding work group (AKSKD), together with raw material producers and research institutes involved in this field.

Because structural adhesives are used for a wide variety of applications in many different industries (for example, car, train and aircraft production and shipbuilding, the electrical and electronics industry, the household appliances industry, medical technology, the optical industry, mechanical engineering, plant construction and equipment manufacturing, wind power and solar energy), the committee covers technical subjects of general interest and those relating to specific market segments.

The main areas of the committee's work include:
- **Training in adhesive technology:** Thorough training for users of adhesives and the resulting clear understanding of its applications help to ensure that adhesives are used successfully, correctly and in accordance with the requirements of specific production processes. The training is therefore in the interests of adhesive manufacturers. This is reinforced by the DIN 2304-1 standard, which was published in early 2016. It requires employees who take responsibility for planning and implementing safety-related structural bonds to have the appropriate qualifications. The committee supports the three-stage DVS/EWF training concept, which has also gained international acceptance, and the preparation of the guidelines "Kleben – aber richtig" ("Adhesive Bonding – the Right Way"), which are available on the IVK website.

The restrictions caused by the COVID-19 pandemic have led to a fall in the number of participants at the two training centres, but also to the development of new concepts, such as blended learning, which is a combination of web-based self study and classroom teaching

- **Promoting research:** TKSKD also sees itself as a bridge between industry and scientific activities that take place outside the industrial world and regularly provides information about topics relevant to structural adhesives as part of planned and approved pre-competitive adhesive research projects. Where a need for research is identified, TKSKD works closely with research bodies to submit applications for new projects. After the projects have been approved, it plays an active role on project monitoring committees.

For instance, the project run by the Fraunhofer Institute for Manufacturing Technology and Advanced Materials IFAM and the SKZ on the subject of the wetting envelope was supported by individual member companies. The latest status of the project has been reported to AKSKD. The goal of the project is to transfer the process, which has been successfully used in the application of coatings, to adhesives and to allow for more accurate predictions of the wetting behaviour of surfaces (bonded parts) when liquids (adhesives) are applied to them and therefore to enable the development of adhesive forces. As many of the properties of adhesives differ from those of conventional coating materials, the existing methods for determining the required parameters must be tested for their applicability to adhesives and, if necessary, modified.

- **Standards activities:** The experts from IVK member companies who are members of national and international standards committees report regularly on activities relating to structural adhesives in the following working groups:
  - CEN/TC 193   Adhesives
  - ISO/TC 61/SC 11/WG5   Polymeric adhesives
  - DIN NA 062-10-02 AA   Test methods relevant to structural adhesive technology
  - NA 062-10-03 AA   Adhesive bondings in electronic applications
  - DIN NA 062-10 FBR   Board of the adhesives technical department
  - DIN NA 087-05-06 AA   Adhesive bonding technology in rail vehicle construction
  - DIN/DVS NAS 092-00-28 AA   Adhesive bonding committee
  - DVS AG V 8   Adhesive bonding
  - DVS AG W 4.14   Joining CFRP

- The topics covered over the past few years include:
  - Information on the current status of the internationalisation of Germany's rail vehicle standard DIN 6701 Adhesive bonding of railway vehicles and parts. This standard defines the mandatory requirements placed on companies that manufacture bonded rail vehicles or parts for use on Germany's railways. It is now being turned into a European standard.
  - Information on the work of the standards committee "Quality assurance of adhesive bonds". TKSKD members on this committee have been actively involved in drawing up DIN standard 2304-1 *Adhesive bonding technology – Quality requirements for adhesive bonding processes – Part 1: Adhesive bonding process chain and the documents* that provide more detail on this standard.
    - DIN SPEC 2305-1 *Adhesive bonding technology – Process chain adhesive bonding – Part 1: Advice for manufacturing*
    - DIN SPEC 2305-2 *Adhesive bonding technology – Quality requirements for adhesive bonding processes – Part 2: Adhesive bonding of fibre composite materials* Summary of the special factors involved in bonding fibre composites and the resulting requirements for the bonding process chain.
    - DIN SPEC 2305-3 *Adhesive bonding technology – Quality requirements for adhesive bonding processes – Part 3: Requirements for the adhesive bonding personnel* Additional information about the requirements for employees in accordance with section 5.2 of DIN 2304-1 for bonds in safety classes S1 to S3.

The experts from IVK member companies who sit on the standards committees provide the AKSKD members with regular information about the status of the standards.

- **Chemicals legislation:** As in previous reporting periods, the latest developments in chemicals legislation and the resulting requirements for the manufacturers of adhesives and their customers remained a key topic. Examples of these include:
  - New inclusions in the list of substances of very high concern (SVHCs).
  - The status of the planned restrictions on the use of diisocyanates. The German Federal Institute for Occupational Safety and Health (BAuA) has initiated a REACH restriction process under the terms of Annex XV for this group of substances with the aim of avoiding the respiratory sensitivities caused by diisocyanates and improving occupational health and safety in professional and industrial environments. A proposal was submitted to the European Chemicals Agency (ECHA) for this purpose in October 2016. The restriction is not intended to lead to a general ban on the chemicals. Instead, the plan is to introduce mandatory, verifiable requirements for protective measures and training courses in the safe handling of diisocyanates. The training materials are currently being prepared by the manufacturers of diisocyanates with the involvement of FEICA.
  - The creation of a harmonised European occupational exposure limit (OEL) for diisocyanates. Depending on how low this limit is, it may require a measurement method with a much lower detection limit, which could in turn lead to the need for a considerable investment in measuring instruments. In addition, a very low limit could also result in both adhesive manufacturers and their customers having to invest in other equipment (such as extraction systems).
  - The process for reporting to national poison centres any adhesive formulations that have been classified as harmful to health and the implementation of the process.
  - Registration of polymers in the context of REACH. Currently REACH requires monomers but not the resulting polymers to be registered. Because of recent developments (including the plastics strategy and microplastics), it now seems certain that the European Commission will present a legislative proposal for the registration of polymers.
  - The circular economy and the EU Green Deal. During the reporting period, information on these subjects and the resulting issues were of considerable importance.

Public Relations Advisory Board (BeifÖ)

The main task of the Public Relations Advisory Board is to present the German Adhesives Association (IVK) and the key technology of adhesive bonding in all its depth and complexity to the general public and the media in a positive light. In November 2019, Thorsten Krimphove (the public relations officer of WEICON) was chosen as the new spokesperson of the advisory board. He succeeds Ulrich Lipper (former managing director of Cyberbond), who has retired.

The ongoing public relations work of the German Adhesives Association has proved to be a great success. The subject of adhesive bonding is now firmly established in the print and online media, and the media response is correspondingly high. IVK's press and public relations work has an average annual circulation of around 120 million copies.

As the old IVK website was no longer adequate in either technical or structural terms for the large amount of information it contained, which had been constantly increasing in recent years, the website was completely overhauled in the first half of 2020. The Public Relations Advisory Board played an important role in this process. You will now find all the information about the association and the world of adhesive bonding clearly presented at www.klebstoffe.com. The improved navigation system and the new structure of the content make access to the information more intuitive and faster. In addition, the IVK online magazine "Kleben fürs Leben" (Bonding for Life) has been coordinated with and integrated into the website. The site has been designed for all the current browsers and mobile devices, including tablets, and meets the latest standards. Information and downloads are still available exclusively to members of the association in a separate intranet area.

IVK also has a presence on the relevant social media platforms, including Facebook, LinkedIn, Twitter and YouTube. Adhesive bonding is a theme across all the social media channels.

The print magazine "Kleben fürs Leben" (Bonding for Life) has become a permanent feature of the association's PR activities. Published once a year, it forms an essential component of the communication strategy of the German adhesives industry and plays a key role in promoting the industry's activities. The magazine is free from any form of product or corporate advertising and focuses on further strengthening the positive image of the adhesives industry and documenting the wide-ranging benefits of bonding technologies. This year's issue covers the 75th anniversary of the German Adhesives Association and, as always, highlights applications for adhesives from a variety of areas of our daily lives. The magazine has been produced on recycled paper using sustainable printing and bonding processes. "Kleben fürs Leben" will be published in 2021 for the thirteenth time.

The IVK promotional video "Faszination Kleben" (The Fascination of Bonding) demonstrates that adhesives are an essential feature of everyday life in the home, in the construction trades and in industry. It also explains why many technologies of the future and current production processes for everyday objects are only made possible by the use of adhesives. In the space of five minutes, viewers not only discover important facts about the chemical aspects of adhesives and how they function, but also find out about the different areas where adhesives are successfully used. Almost every industry, from the automotive sector to electrical engineering, textiles and clothing, relies on bonding to improve the quality of its products and introduce innovations. Another video entitled "Was unsere Welt zusammenhält" (What Holds Our World Together) has been produced in cooperation with the Association of the European Adhesive and Sealant Industry (FEICA).

Against the background of the German Federal Government's digitisation strategy, the association has worked together with the German Chemical Industry Fund to develop digital educational materials. These innovative learning tools proved popular with teachers and parents for use as part of home schooling. A range of interactive infographics designed specifically for use on whiteboards, PCs and tablets in schools has been created to explain the subject of adhesives to school students using modern teaching methods.

In addition, the association has commissioned the Institute of the Didactics of Chemistry at the University of Frankfurt to develop a teacher training programme on the subject of "Adhesives and bonding in teaching" in order to give a more in-depth insight into the chemistry of adhesives and to highlight links with the school curriculum.

The new material was completed in time for the start of the 2018/2019 academic year and is available free of charge to teachers throughout Germany.

Management

The employees of the German Adhesives Association's office are responsible for coordinating, handling and following up on the wide variety of tasks which come from various committees. The office keeps members up to date on new topics that are important for the industry and its specialised committees. It also serves as an information exchange and a point of contact for the association's members with regard to relevant technical and legal information relating to health and safety at work, environmental and consumer protection and to legislation covering areas such as competition, chemicals, the environment, food and sustainability.

The Management Board of IVK perceives itself as the representative and expert partner of the adhesives industry. In this role, the Management Board represents the technical and economic interests of the industry among German, European and international institutions. In addition, the employees of the association maintain a close and proactive dialogue with customer, trade and consumer associations, system partners, scientific institutions and the general public. Through active involvement in the advisory boards of major industry exhibitions and trade journals, as well as the working committees of various German and European ministries, the Management Board uses its knowledge and expertise to monitor and support adhesives-related topics and projects on all levels and throughout the entire adhesives value chain.

This also applies to research. In its role as a member of the advisory board of the ProcessNet Adhesive Technology division and the joint committee on adhesive bonding (GAK), the Management Board helps promote scientific research and coordinate publicly funded scientific research projects in the field of adhesive bonding technology.

As part of a broad portfolio of publications, discussions with interested groups and specialist lectures, the Management Board effectively communicates the adhesives industry's broad range of services and extensive potential for innovation as well as the exemplary commitment of its members to protecting workers, consumers and the environment.

With effect from 1 January 2021, the IVK board member Dr Vera Haye has been appointed the new managing director with overall responsibility for the German Adhesives Association. Vera Haye (41) has a doctorate in microbiology and succeeds Ansgar van Halteren, who is retiring after working for the association for 38 years.

PERSONAL DATA

Honorary members and honorary chair

At the general assembly in 2012, Arnd Picker was appointed an honorary member of the German Adhesives Association and, at the same time, its honorary chair. In this way, the association and its members recognised the achievements of Arnd Picker, who ensured that the association was in a strong position during his 16 years as chair of the Executive Board. The German Adhesives Association is the world's largest and, in terms of its comprehensive portfolio of services for members, the world's leading association for adhesive bonding technology.

Honorary Members
Arnd Picker, Honorary chair
Dr. Johannes Dahs
Dr. Hannes Frank
Dr. Rainer Vogel

Achievement Medal of the German Adhesives Industry

The German Adhesives Association awards the Achievement Medal of the German Adhesives Industry to individuals for their outstanding service to the adhesives industry and adhesives technology.

The medal was awarded to

Peter Rambusch – May 2018

for the significant contribution he has made to breaking down the organisational barriers between manufacturers of adhesives and producers of self-adhesive tapes which existed until the 1990s. On his initiative, commercial and technical representatives of German adhesive tape manufacturers were admitted into the German Adhesives Association in 1995. Peter Rambusch has continued to represent the interests of the adhesive tape industry on the Executive Board of the German Adhesives Association in a competent, effective and committed way up to the present day. He has not only played an outstanding role in promoting cooperation between two closely related industries, but, at the same time, has also significantly broadened the skills profile of the German Adhesives Association with regard to all technical adhesive issues.

Marlene Doobe – June 2017

for her decades of commitment to the German adhesives industry. As editor-in-chief of the magazine adhäsion KLEBEN + DICHTEN, she has successfully led this important trade publication for the adhesives industry for more than 20 years, with professionalism, the highest level of personal dedication and a great deal of passion. During this time and in close cooperation with the German Adhesives Association, she has developed the magazine into an interdisciplinary, 360-degree communication platform for adhesives technology. Marlene Doobe has made an extremely valuable contribution to promoting close cooperation between research and industry in the field of adhesives technology and has advanced and shaped key aspects of the adhesives industry.

Dr. Manfred Dollhausen – May 2010

in recognition of his successful involvement in standardisation, which made a major contribution to documenting and representing adhesive technology "made in Germany" in national, European and international standards. In addition to that, as early as the 1960s Dr Dollhausen recognised the invaluable significance of technical cooperation between the adhesives industry and the raw material industry. He actively promoted this cooperation and laid the foundations for the partnership between the two industries that is still successful to this day.

Dr. Hannes Frank – September 2007

in recognition of his many years of dedicated service to the German adhesives industry. As a member of the Technical Board, he promoted and helped shape both adhesive technology and the image of the adhesives industry. This includes in particular his commitment to small and medium-sized enterprises and their potential to innovate, which is essential for technical and economic development. Dr. Frank is also regarded as a successful pioneer in the field of polyurethane adhesive technology. Furthermore, he has promoted an industry-wide strategy on communication and training, which has helped to establish adhesives as a key technology of the 21st century.

Prof. Dr. Otto-Diedrich Hennemann – May 2007

in recognition of his scientific work which promoted and helped shape the system of adhesive bonding. This includes his research into the durability of bonds and the implementation of appropriate simulation processes in the automotive and aerospace industries. His approach to research was to always focus on practical applications and on creating added value for partners.

Committees of the German Adhesives Association (IVK)

You can find information about the current structure and membership of the committees at www.klebstoffe.com/organisation-und-struktur/.

Executive Board

Chair: Dr. Boris Tasche	Henkel AG & Co. KGaA D-40191 Düsseldorf
Deputy chair: Dr. René Rambusch	certoplast Technische Klebebänder GmbH D-42285 Wuppertal
Other members:	
Dirk Brandenburger	Sika Automotive Hamburg GmbH D-22525 Hamburg
Mark Eslamlooy	ARDEX GmbH D-58453 Witten
Stephan Frischmuth	tesa SE D-22848 Norderstedt
Dr. Gert Heckmann	Kömmerling Chemische Fabrik GmbH D-66954 Pirmasens
Dr. Hans-Georg Kinzelmann	Henkel AG & Co. KGaA D-40191 Düsseldorf
Timm Koepchen	EUKALIN Spezial-Klebstoff Fabrik GmbH D-52249 Eschweiler
Klaus Kullmann	Jowat SE D-32758 Detmold
Olaf Memmen	Bostik GmbH D-33829 Borgholzhausen

Dr. Thomas Pfeiffer	Türmerleim GmbH D-67061 Ludwigshafen
Dr. Christoph Riemer	Wacker Chemie AG D-81737 München
Leonhard Ritzhaupt	Kleiberit Klebchemie M. G. Becker GmbH & Co. KG D-76356 Weingarten
Philipp Utz	UZIN UTZ AG D-89079 Ulm

Technical Board

| Chair:
Dr. Hans-Georg Kinzelmann | Henkel AG & Co. KGaA
D-40191 Düsseldorf |

Other members:

Dr. Norbert Arnold	UZIN UTZ AG D-89079 Ulm
Dr. Rainer Buchholz	RENIA Ges. mbH chemische Fabrik D-51076 Köln
Dr. Torsten Funk	Sika Technology AG CH–8048 Zürich
Seda Gellings	UHU GmbH & Co. KG D-77815 Bühl
Prof. Dr. Andreas Groß	Fraunhofer-Institut für Fertigungstechnik und Angewandte Materialforschung (IFAM) D-28359 Bremen
Daniela Hardt	Celanese Services Germany GmbH D-65844 Sulzbach
Christoph Küsters	3M Deutschland GmbH D-41453 Neuss

Dr. Annett Linemann	H.B. Fuller Deutschland GmbH D-21335 Lüneburg
Dr. Hartwig Lohse	Klebtechnik Dr. Hartwig Lohse e. K. D-25524 Itzehoe
Dr. Michael Nitsche	Bostik GmbH D-33829 Borgholzhausen
Matthias Pfeiffer	Türmerleim GmbH D-67061 Ludwigshafen
Arno Prumbach	EUKALIN Spezial-Klebstoff Fabrik GmbH D-52249 Eschweiler
Leonhard Ritzhaupt	Kleiberit Klebchemie M. G. Becker GmbH & Co. KG D-76356 Weingarten
Dr. Karsten Seitz	tesa SE D-22848 Norderstedt
Dr. Christian Terfloth	Jowat SE D-32758 Detmold
Dr. Christoph Thiebes	Covestro Deutschland AG D-51365 Leverkusen
Dr. Axel Weiss	BASF SE D-67063 Ludwigshafen

Technical Committee Building Adhesives

Chair: Dr. Norbert Arnold	UZIN UTZ AG D-89079 Ulm
Other members:	
Dr. Thomas Brokamp	Bona GmbH Deutschland D-65549 Limburg

Dr. Michael Erberich	WULFF GmbH & Co. KG D-49504 Lotte
Manfred Friedrich	Sika Deutschland GmbH D-48720 Rosendahl
Dr. Frank Gahlmann	Stauf Klebstoffwerk GmbH D-57234 Wilnsdorf
Stefan Großmann	Sopro Bauchemie GmbH D-65203 Wiesbaden
Dr. Matthias Hirsch	Kiesel Bauchemie GmbH & Co. KG D-73730 Esslingen
Michael Illing	Forbo Eurocol Deutschland GmbH D-99091 Erfurt
Christopher Kupka	Celanese Services Germany GmbH D-65843 Sulzbach
Bernd Lesker	Mapei GmbH D-46236 Bottrop
Dr. Michael Müller	Bostik GmbH D-33829 Borgholzhausen
Dr. Maximilian Rüllmann	BASF SE D-67063 Ludwigshafen
Dr. Martin Schäfer	Wakol GmbH D-66954 Pirmasens
Michael Schäfer	BCD Chemie GmbH D-21079 Hamburg
Dr. Jörg Sieksmeier	ARDEX GmbH D-58453 Witten

Hartmut Urbath	PCI Augsburg GmbH D-59071 Hamm
Dr. Steffen Wunderlich	Kleiberit Klebstoffe Klebchemie M. G. Becker GmbH & Co. KG D-76356 Weingarten

Technical Committee on Wood Adhesives

Chair: Daniela Hardt	Celanese Services Germany GmbH D-65844 Sulzbach

Other members:

Wolfgang Arndt	Covestro Deutschland AG D-51365 Leverkusen
Holger Brandt	Follmann & Co. GmbH & Co. KG D-32423 Minden
Christoph Funke	Jowat SE D-32758 Detmold
Oliver Hartz	BASF SE D-67063 Ludwigshafen
Dr. Thomas Kotre	Planatol GmbH D-83101 Rohrdorf-Thansau
Jürgen Lotz	Henkel AG & Co. KGaA D-73442 Bopfingen
Dr. Marcel Ruppert	Wacker Chemie AG D-84489 Burghausen
Martin Sauerland	H.B. Fuller Deutschland GmbH D-21335 Lüneburg
Holger Scherrenbacher	Kleiberit Klebstoffe Klebchemie M. G. Becker GmbH & Co. KG D-76356 Weingarten

Technical Committee DIY and Consumer Adhesives

Chair: Seda Gellings	UHU GmbH & Co. KG D-77815 Bühl

Other members:

Frank Avemaria	3M Deutschland GmbH D-41453 Neuss
Dr. Nils Hellwig	Henkel AG & Co. KGaA D-40191 Düsseldorf
Dr. Florian Kopp	RUDERER KLEBETECHNIK GMBH NL-85604 Zorneding
Henning Voß	WEICON GmbH & Co. KG D-48157 Münster

Technical Committee Adhesive Tapes

Chair: Dr. Karsten Seitz	tesa SE D-22848 Norderstedt

Other members:

Dr. Achim Böhme	3M Deutschland GmbH D-41453 Neuss
Dr. Thomas Christ	BASF SE D-67063 Ludwigshafen
Dr. Ruben Friedland	Lohmann GmbH & Co. KG D-56567 Neuwied
Dr. Thomas Hanhörster	Sika Automotive GmbH D-22525 Hamburg
Prof. Dr. Andreas Hartwig	Fraunhofer-Institut für Fertigungstechnik und Angewandte Materialforschung (IFAM) D-28359 Bremen

Lutz Jacob	RJ Consulting GbR D-87527 Altstaedten
Dr. Thorsten Meier	certoplast Technische Klebebänder GmbH D-42285 Wuppertal
Melanie Ott	H.B. Fuller Deutschland GmbH D-21335 Lüneburg
Dr. Ralf Rönisch	COROPLAST Fritz Müller GmbH & Co. KG D-42279 Wuppertal
Dr. Jürgen K. L. Schneider	TSRC (Lux.) Corporation S.a.r.l. L-1931 Luxemburg
Michael Schürmann	Henkel AG & Co. KGaA D-40191 Düsseldorf

Technical Committee Paper and Packaging Adhesives

Chair: Arno Prumbach	EUKALIN Spezial-Klebstoff Fabrik GmbH D-52249 Eschweiler

Other members:

Dr. Elke Andresen	Bostik GmbH D-33829 Borgholzhausen
Dr. Olga Dulachyk	Gludan (Deutschland) GmbH D-21514 Büchen
Holger Hartmann	Celanese Services Germany GmbH D-65843 Sulzbach
Dr. Gerhard Koegler	Wacker Chemie AG D-84489 Burghausen
Dr. Thomas Kotre	Planatol GmbH D-83101 Rohrdorf-Thansau
Matthias Pfeiffer	Türmerleim GmbH D-67061 Ludwigshafen

Dr. Peter Preishuber-Pflügl	BASF SE D-67063 Ludwigshafen
Alexandra Ross	H.B. Fuller Deutschland GmbH D-21335 Lüneburg
Michael Schäfer	BCD Chemie GmbH D-21079 Hamburg
Dr. Christian Schmidt	Jowat SE D-32758 Detmold
Julia Szincsak	Follmann Chemie GmbH D-32423 Minden
Dr. Monika Tönnießen	Henkel AG & Co. KGaA D-40191 Düsseldorf

Technical Committee Shoe Adhesives

Chair: Dr. Rainer Buchholz	RENIA Ges. mbH chemische Fabrik D-51076 Köln
Other members:	
Wolfgang Arndt	Covestro Deutschland AG D-51368 Leverkusen
Martin Breiner	Kömmerling Chemische Fabrik GmbH D-66929 Pirmasens
Andreas Ecker	H.B. FULLER Austria GmbH A-4600 Wels
Dr. Martin Schneider	ARLANXEO Deutschland GmbH D-41538 Dormagen

Technical Committee Structural Adhesives and Sealants

Chair: Dr. Hartwig Lohse	Klebtechnik Dr. Hartwig Lohse e. K. D-25524 Itzehoe

Other members:

Dr. Beate Baumbach	Covestro Deutschland AG D-51368 Leverkusen
Jürgen Fritz	Evonik Resource Efficiency GmbH D-79618 Rheinfelden
Dr. Oliver Glosch	Weiss Chemie + Technik GmbH & Co. KG D-35703 Haiger
Rosto Iosif	3M Deutschland GmbH D-41453 Neuss
Dr. Stefan Kreiling	Henkel AG & Co. KGaA Heidelberg Site D-69112 Heidelberg
Dr. Erik Meiß	Fraunhofer-Institut für Fertigungstechnik und Angewandte Materialforschung (IFAM) D-28359 Bremen
Michael Schäfer	BCD Chemie GmbH D-21079 Hamburg
Bernhard Schuck	Bostik GmbH D-33829 Borgholzhausen
Frank Steegmanns	Stockmeier Urethanes GmbH & Co. KG D-32657 Lemgo
Artur Zanotti	Sika Deutschland GmbH D-72574 Bad Urach

Advisory Board for Sustainability

Spokesperson: Jürgen Germann	3M Deutschland GmbH D-41453 Neuss
Deputy spokesperson: Dr. Peter Krüger	Covestro Deutschland AG D-51373 Leverkusen
Other members:	
Dr. Norbert Arnold	UZIN UTZ AG D-89079 Ulm
Dr. Jörg Dietrich	POLY-CHEM GmbH D-06766 Bitterfeld-Wolfen
Uwe Düsterwald	BASF SE D-67063 Ludwigshafen
Dr. Ruben Friedland	Lohmann GmbH & Co. KG D-56567 Neuwied
Prof. Dr. Andreas Hartwig	Fraunhofer Institute for Manufacturing Technology and Advanced Materials (IFAM) D-28359 Bremen
Ulla Hüppe	Henkel AG & Co. KGaA D-40191 Düsseldorf
Linn Mehnert	Wacker Chemie AG D-84489 Burghausen
Dr. Thomas Schubert	tesa SE D-22848 Norderstedt
Timm Schulze	Jowat SE D-32758 Detmold

Marcus Wedemann	KRAHN CHEMIE GMBH D-20457 Hamburg
Dr. Martin Weller	H.B. Fuller Deutschland GmbH D-21335 Lüneburg

Public Relations Advisory Board

Spokesperson: Thorsten Krimphove	WEICON GmbH & Co. KG D-48157 Münster

Other members:

Tamara Beiler	Innotech Marketing und Konfektion Rot GmbH D-69242 Rettigheim
Holger Bleich	Cyberbond Europe GmbH A H.B. Fuller Company D-31515 Wunstorf
Qiyong (James) Cui	Wacker Chemie AG D-81737 München
Sebastian Hinz	Henkel AG & Co. KGaA D-40191 Düsseldorf
Oliver Jüntgen	Henkel AG & Co. KGaA D-40191 Düsseldorf
Jens Ruderer	RUDERER KLEBETECHNIK GMBH D-85600 Zorneding
Jan Schulz-Wachler	EUKALIN Spezial-Klebstoff Fabrik GmbH D-52249 Eschweiler
Dr. Christine Wagner	Wacker Chemie AG D-84489 Burghausen

Work Group Building Adhesives

Chair:
Olaf Memmen

Bostik GmbH
D-33829 Borgholzhausen

Work Group Wood Adhesives

Chair:
Klaus Kullmann

Jowat SE
D-32758 Detmold

Work Group Industrial Adhesives

Chair:
Dr. Boris Tasche

Henkel AG & Co. KGaA
D-40191 Düsseldorf

Work Group Adhesive Tapes

Chair:
Dr. René Rambusch

certoplast Technische Klebebänder GmbH
D-42285 Wuppertal

Work Group Paper/Packaging Adhesives

Chair:
Dr. Thomas Pfeiffer

Türmerleim GmbH
D-67014 Ludwigshafen

Work Group Raw Materials

Chair:
Dr. Christoph Riemer

Wacker Chemie AG
D-81737 München

Work Group Structural Adhesives and Sealants

Chair:
Dirk Brandenburger

Sika Automotive Hamburg GmbH
D-22525 Hamburg

Work Group DIY and Consumer Adhesives

Chair:
Seda Gellings

UHU GmbH & Co. KG
D-77815 Bühl

Work Group Foam Adhesives

Chair:
Norbert Uniatowsky

Bostik GmbH
D-33829 Borgholzhausen

Work Group Shoe Adhesives

Chair:
Dr. Rainer Buchholz

RENIA Ges. mbH chemische Fabrik
D-51076 Köln

Management Board

Dr. Vera Haye	Managing Director
Dr. Axel Heßland	Technology Director
Klaus Winkels	Legal Director
Michaela Szkudlarek	Assistant to the Managing Director/Finance
Danuta Dworaczek	Consultant for Technology and Research
Nathalie Schlößer	Public Relations Manager
Martina Weinberg	Events Manager
Natascha Zapolowski	Consultant for Environment and Technology

Honorary Chair

Arnd Picker

Rommerskirchen

Honorary Members

Dr. Johannes Dahs	Königswinter
Dr. Hannes Frank	Detmold
Arnd Picker	Rommerskirchen
Dr. Rainer Vogel	Langenfeld

Holders of the Achievement Medal of the German Adhesives Industry

Peter Rambusch	Wuppertal
Marlene Doobe	Eltville
Dr. Manfred Dollhausen	Odenthal
Dr. Hannes Frank	Detmold
Prof. Dr. Otto-D. Hennemann	Osterholz-Scharmbeck

REPORT 2020/2021

FCIO – Austria

FCIO – Austria

The Construction Adhesives Professional Group of the Association of the Austrian Chemical Industry was founded in 2008 as the successor to the disbanded Association of Austrian Adhesives Manufacturers (VÖK). Within the Association of the Austrian Chemical Industry (FCIO), the Construction Adhesives Professional Group functions as an independent organisation.

The Construction Adhesives Professional Group currently has ten members.

Mission and services

The FCIO Construction Adhesives Professional Group with its ten members is a group based on the Austrian Chamber of Commerce Act within the Association of the Austrian Chemical Industry. The main task of the group is to represent the interests of its members and help shape the economic conditions for the industry in Austria. As an entity under public law, the Construction Adhesives Professional Group has the legal mandate to protect the interests of the industry in all areas and to advise its member companies on all legal matters, in particular those relating to environmental and employment law. The group is in permanent contact with the relevant public bodies and trades unions. It is also part of several working groups organised by scientific institutions and ministries and is a member of national standards committees. The group is responsible for advising its members and helping them to meet their legal obligations, in particular with regard to health and safety and environmental protection. The group is also involved in training programmes for apprentices in the tiling and floor laying trade in Austria.

Organisation and structure

The Construction Adhesives Professional Group is part of the Association of the Austrian Chemical Industry (FCIO), which in turn comes under the umbrella of the Austrian Chamber of Commerce organisation (WKO).

President: Bernhard Mucherl/Murexin AG
Managing director: Dr Klaus Schaubmayr/FCIO

REPORT 2020/2021

FKS – Switzerland

FKS – Switzerland

Association of the Adhesives Industry Switzerland (FKS)

Our role

The association supports its members with regard to adhesive and sealant production, in particular through:

- Representing the interests of the Swiss adhesives and sealants industry in public bodies and associations, including involvement in legislative activities
- Taking part in technical committees to improve the industry's cooperation with public bodies and national and international associations
- Supplying statistics and basic information about the adhesives market in Switzerland to Swiss and European public bodies for use in their decision-making processes
- Providing technical information and expertise to promote the trust of customers in the members of the association
- Encouraging members to share information and experiences regularly with the aim of improving the quality of the industry's products even further
- Organising technical lectures and presentations

Market developments, regulations and measures

The association monitors market developments and existing regulations as the basis for the implementation of measures relating to environmental protection and safety during the production, packaging, transport, use and disposal of adhesives. These measures and the association's services help to ensure that the most demanding market requirements can be met at all times.

Member of FEICA

(Association of the European Adhesive and Sealant Industry)

The association is a member of FEICA, which represents the interests of the national associations on an international level in cooperation with international organisations. FEICA regularly provides the national associations with information about developments in Europe.

Service

- Statistics and basic information
- National standards
- Technical information and expertise
- Sharing information and experiences
- Information about regulatory changes
- Spring and autumn conferences
- Access for members to FKS information in a separate area of the website

Organisation and structure
- President
  Toni Rüegg
- Vice-President
  Andreas Mosimann
- Secretary
  Simon Bienz

Members

Alfa Klebstoffe AG	H.B. Fuller Europe GmbH
APM Technica AG	Henkel & Cie. AG
Artimelt AG	JOWAT Swiss AG
ASTORtec AG	Kisling AG
Avery Dennison Materials Europe GmbH	KVT-Fastening, a Bossard AG company
BFH Architektur, Holz und Bau	merz+benteli ag
Collano AG	nolax AG
Distona AG	Pontacol AG
DuPont Transportation & Industrial	Sika Schweiz AG
Emerell AG	Türmerleim AG
EMS-CHEMIE AG	Uzin Utz Schweiz AG
ETH Zurich	Wakol GmbH
FHNW – University of Applied Sciences Northwestern Switzerland	ZHAW – Zurich University of Applied Sciences
GYSO AG	

Contact Information

President	Vice-President	Secretary	Secretariat
Toni Rüegg	Andreas Mosimann	Simon Bienz	Association of the Adhesives
JOWAT Swiss AG	Sika Schweiz AG	merz+benteli AG	Industry Switzerland
Schiltwaldstrasse 33	Tüffenwies 16	Freiburgstrasse 616	Silvia Fasel
CH-6033 Buchrain	CH-8048 Zurich	CH-3172 Niederwangen	Bahnhofplatz 2 a
Phone: +41 (0) 41 445 11 11	Phone: +41 (0) 58 436 40 40	Phone: +41 (0) 31 980 48 48	CH-5400 Baden
Fax: +41 (0) 41 440 23 46	Email: mosimann.andreas@	Fax: +41 (0) 31 980 48 49	Phone: +41 (0) 56 221 51 00
Email:	ch.sika.com	Email:	Fax: +41 (0) 56 221 51 41
toni.rueegg@jowat.ch		simon.bienz@merz-benteli.ch	E-Mail: info@fks.ch
			www.fks.ch

REPORT 2020/2021

VLK – Netherland

VLK – Netherland

VLK

Industrial Association

The Vereniging Lijmen en Kitten (VLK) represents the adhesives and sealants industry in the Netherlands. Because of its technical and economic interests, it is an important player on the European market.

The adhesives and sealants industry in the Netherlands focuses primarily on business to business sales. Adhesives and sealants are mainly used in industry and in the construction sector. It is estimated that the VLK represents the companies responsible for around 75% of the adhesives and sealants sold in the Netherlands..

For the member companies the VLK is:
- *A point of contact* for the general public and public bodies, including inspection and regulatory organisations, institutions of civil society, consumers and other players
- *A spokesperson* for the industry supporting practical legislation on a European level
- *A source of information* and a helpdesk for legislation about materials, such as REACH and the CLP Regulation, and for construction issues, including the CPR and CE marking
- *An organisation that protects* the image of adhesives and sealants and of the adhesives and sealants industry
- *A facilitator* for knowledge sharing and networking.

The VLK is a member of FEICA (the Association of the European Adhesive & Sealant Industry) which represents the industry on a European level.

Services
What can members of the VLK expect?

- Lobbying
The VLK works to ensure that legislation and regulations that affect the development, production and sale of adhesives and sealants in the Netherlands are practical and realistic. The industrial organisation represents the interests of the sector through its contacts with the general public and the institutions of civil society. Many new developments are the results of European legislation, which is why the VLK is a member of FEICA.

- Networking
The VLK is at the centre of a network of companies and plays an important role in bringing them together. Its activities go beyond the adhesives and sealants industry. The VLK maintains contacts with other companies in the supply chain and with academic institutions.

• Knowledge sharing
The VLK identifies relevant information and makes it available to its members via the website and its online newsletters. Member companies also have access to specially developed tutorials on the VLK website. They can share information about developments in legislation and regulations, health and safety, the environment and standards.

• Insights into changes in the market
The flooring adhesives and tile adhesives departments have set benchmarks for developments in the markets that they are involved in. Member companies receive a quarterly report about the most recent activities in the market.

• Helpdesk
The VLK has a helpdesk for issues relating to legislation and regulations that have an impact on adhesives and sealants. Examples include questions about REACH, the CLP Regulation and CE marking. The helpdesk covers both European and Dutch legislation and regulations.

• Positive image and trust
The coordinating role played by the VLK makes it the ideal body to provide information about the adhesives and sealants industry. It helps to ensure the conformity of the industry.

Members
The member companies are the most important feature of the VLK. The industrial department is made up of directors, managers and experts who work on behalf of the VLK central office. The contact details of all the member companies are available on the website at www.vlk.nu/leden. The VLK is not involved in the commercial activities of individual companies.

Organisation
The members of the management board of the VLK are as follows:
• Wybren de Zwart (Saba Dinxperlo BV) – chair
• Rob de Kruijff (Sika Nederland BV)
• Gertjan van Dinther (Soudal BV)
• Gerrit Jonker (Omnicol BV)
• Dirk Breeuwer (Forbo Eurocol)

The VLK consists of the following departments:
• Floor adhesives and levelling compounds
• Tile adhesives
• Sealants

The VLK consists of the following working groups:
• Committee for hazardous materials
• Technical committee for sealants
• Technical committee for tile adhesives
• Technical committee for wall cladding and facade panels

Office

In just the same way as paints and printing inks, adhesives and sealants are mixtures. The VLK works closely with the Dutch association for the paints and printing inks industry (VVVF). It has an office within the VVVF and takes part in the association's industry-wide working groups and meetings.

More information

If you need more information, you can contact the VLK via www.vlk.nu.

VLK

Loire 150, 2491 AK, The Hague, The Netherlands
Phone: + 31 70 444 06 80
Email: info@vlk.nu
www.vlk.nu

GEV

Association for the Control of Emissions
from Products for Flooring Installation,
Adhesives & Building Materials

EMICODE®

Protection from indoor air contaminants

The need to protect people from indoor air contaminants and calls from users for the provision of low-emission products resulted in the foundation of the Association for the Control of Emissions in Products for Flooring Installation, Adhesives and Construction Products (GEV) and the establishment of the EMICODE® labelling system in February 1997. This initiative was supported by leading manufacturers of flooring installation products which are also members of the German Adhesives Association.

The German adhesives industry had originally developed the GISCODE system in the early 1990s in collaboration with the employers' liability insurance organisation for the construction industry, with the aim of helping installers with their choice of flooring installation product. The GISCODE labelling system gives installers a quick overview of products that comply with health and safety legislation.

Under the terms of legislation on hazardous substances, installers of parquet and other types of floor coverings can now only be exposed to high concentrations of volatile solvents in exceptional cases and after taking special protective measures (Technical Rules for Hazardous Sub-

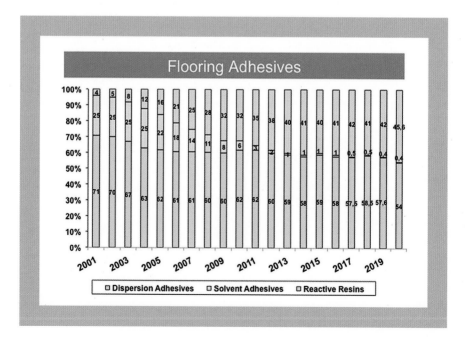

The use of dispersion adhesives

stances (TRGS) 610). The use of substitute substances that represent a much lower risk has become the norm. This has led to a significant number of products containing solvents being phased out over recent years.

Other organic compounds are used in addition to solvents. Some of these are contaminants of raw materials and others are non-volatile compounds. Although they are only present in low concentrations, they can be emitted into the indoor air by the products over long periods.

For this reason, a new generation of solvent-free, very low-emission flooring installation products was developed. These are highly recommended because of their low impact on indoor air quality.

In order to provide processors and consumers with reliable guidance in the light of the large number of different measurement processes that are in use, the objective qualification and labelling system called **EMICODE**® was introduced. This makes it possible to evaluate and compare flooring installation materials and other chemical building products on the basis of their emissions. At the same time, it gives manufacturers a strong incentive to continuously improve their products.

The EMICODE® classes are based on a precisely defined test chamber examination and demanding classification criteria. Adhesives, levelling compounds, precoats, substrates, sealants, fast-drying screeds and other building products which are marked with the GEV label EMICODE® EC 1 to indicate that they are "very low in emissions" produce the lowest possible level of indoor air contaminants. Unlike other systems, with EMICODE® the manufacturers themselves are responsible for labelling, while GEV has samples taken of products on the market by independent institutes for monitoring purposes. Another difference is that GEV does not allow for any compromises on quality. Technically questionable environmental criteria are not permitted for reasons of sustainability.

This voluntary initiative is a systematic continuation of the efforts made to protect the health of processors and consumers. EMICODE® provides contracting authorities, architects, planners, tradespeople, building owners and end consumers with transparent and objective guidelines for selecting low-emission flooring installation products. Videos in 13 languages, brochures, tender document templates, technical documents and the organisation's regulations are available on the website www.emicode.com.

By extending EMICODE® to include products such as joint sealants, parquet varnishes and oils, polyurethane foams, floor screeds, reactive coatings and window films etc., GEV has responded to market requirements for the classification of other products that are not traditional floor installation materials and also to requests from the market and from industry to make it possible to differentiate between products on the basis of their environmental impact.

Chairman of the Executive Board   Stefan Neuberger, Pallmann
Chairman of the Technical Advisory Board   Hartmut Urbath, PCI GmbH
Managing Director of GEV   Klaus Winkels, Attorney-at-law

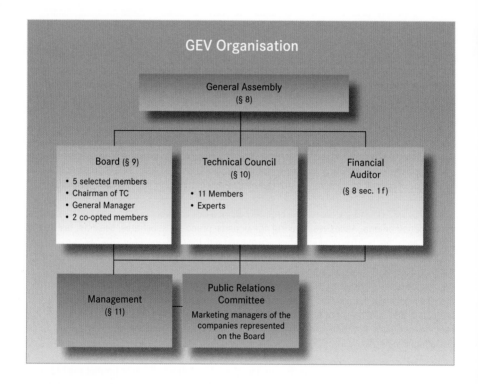

Gemeinschaft Emissionskontrollierte Verlegewerkstoffe,
Klebstoffe und Bauprodukte (GEV)
RWI 4-Haus
Völklinger Straße 4
D-40219 Düsseldorf
Phone +49 (0) 2 11-6 79 31-20
Fax   +49 (0) 2 11-6 79 31-33
Email: info@emicode.com
www.emicode.com

FEICA – Fédération Européenne des Industries de Colles et Adhésifs – has been representing the interests of the European adhesive and sealant industry since 1972. It is the umbrella organisation for 15 national adhesives associations in Europe and it also represents the interests of 24 direct company members, most of which are multinational adhesive manufacturers and producers of assembly foams, plus 19 affiliate company members.

FEICA has its office in Brussels.

Kristel Ons is the Secretary General of the European association.

The objectives of FEICA

Together with its members, FEICA represents the joint interests of the European adhesives industry and the interests of its members, in particular with regard to the institutions of the European Union.

External Relations

To ensure that it obtains information about the plans and regulations of the European institutions (European Parliament, European Council, European Commission, General Directorates) as quickly as possible, FEICA is in close contact with **CEFIC,** the European Chemical Industry Council, and other European associations, many of which are members of the DUCC Group (**D**ownstream **U**ser of **C**hemicals **C**o-ordination Group).

Service for IVK Members

As a national association member of FEICA, the German Adhesives Association represents the interests of German adhesives manufacturers on the Executive Board, the Technical Board and many of the technical committees of the European association. This structure ensures that members of the German Adhesives Association receive an appropriate service free of charge, together with all the necessary information and regular contacts on a European level.

Contact Address

FEICA – Rue Belliard 40 – 1040 Brussels – Belgium – www.feica.eu

EUROPEAN LEGISLATION AND REGULATIONS

Relevant Laws and Regulations
for Adhesives

Hazardous Material Law

- **REACH Regulation** – Regulation (EC) No 1907/2006 of the European Parliament and of the Council of 18 December 2006 concerning the Registration, Evaluation, Authorization and Restriction of Chemicals (REACH), establishing a European Chemicals Agency, amending Directive 1999/45/EC and repealing Council Regulation (EEC) No 793/93 and Commission Regulation (EC) No 1488/94 as well as Council Directive 76/769/EEC and Commission Directives 91/155/EEC, 93/67/EEC, 93/105/EEC and 2000/21/EC
- **CLP Regulation** – Regulation (EC) No 1272/2008 of the European Parliament and of the Council of 16 December 2008 on classification, labelling and packaging of substances and mixtures, amending and repealing Directives 67/548/EEC and 199/45/EC and amending Regulation (EC) No 1907/2006
- **Fees Regulation REACH** – Commission Regulation (EC) No 340/2008 of 16 April 2008 on the fees and charges payable to the European Chemicals Agency pursuant to Regulation (EC) No 1907/2006 of the European Parliament and of the Council on the Registration, Evaluation, Authorization and Restriction of Chemicals (REACH)
- **Fees Regulation CLP** – Commission Regulation (EU) No 44/2010 of 21 May 2010 on the fees payable to the European Chemicals Agency pursuant to Regulation (EC) No 1272/2008 of the European Parliament and of the Council on classification, labelling and packaging of substances and mixtures
- **Chemicals test method Regulation** – Council Regulation (EC) No 440/2008 of 30 May 2008 laying down test methods pursuant to Regulation (EC) No 1907/2006 of the European Parliament and of the Council on the Registration, Evaluation, Authorization and Restriction of Chemicals (REACH)
- **Biocidal Products Regulation** – Regulation (EU) No 528/2012 of the European Parliament and of the Council of 22 May 2012 concerning the making available on the market and use of biocidal products
- **Fees Regulation BPR** – Commission Implementing Regulation (EU) No 564/2013 of 18 June 2013 on the fees and charges payable to the European Chemicals Agency pursuant to Regulation (EU) No 528/2012 of the European Parliament and of the Council concerning the making available on the market and use of biocidal products
- **PIC Regulation** – Regulation (EC) No 649/2012 of the European Parliament and of the Council concerning the export and import of hazardous chemicals (The new text was adopted on 4 July 2012 and will be applicable from 1 March 2014)
- **POP Regulation** – Regulation (EC) No 850/2004 of the European Parliament and of the Council of 29 April 2004 on persistent organic pollutants and amending Directive 79/117/EEC
- **Chemical Agents Directive (CAD)** – Council Directive 98/24/EC of 7 April 1998 on the protection of the health and safety of workers from the risks related to chemical agents at work (fourteenth individual Directive within the meaning of Article 16(1) of Directive 89/391/EEC).
- **Carcinogens and Mutagens Directive (CMD)** - Directive 2004/37/EC of the European Parliament and of the Council of 29 April 2004 on the protection of workers from the risks related to exposure to carcinogens or mutagens at work (sixth individual Directive within the meaning of Article 16(1) of Council Directive 89/391/EEC).

Circular economy

- **The European Green Deal** – Communication from the Commission to the European Parliament, the European Council, the Council, the European Economic and Social Committee and the Committee of the Regions COM/2019/640 final
- **A new Circular Economy Action Plan for a cleaner and more competitive Europe** Communication from The Commission To The European Parliament, The Council, The European Economic And Social Committee And The Committee Of The Regions COM/2020/98 final
- **The EU's chemicals strategy for sustainability towards a toxic-free environment** Communication from the Commission to the European Parliament, the Council, the European Economic and Social Committee and the Committee of the Regions COM(2020) 667 final
- **A European Strategy for Plastics in a Circular Economy** – Communication from the Commission to the European Parliament, the Council, the European Economic and Social Committee and the Committee of the Regions a – COM(2018) 28 final
- **Ecodesign directive** - Directive 2009/125/EC of the European Parliament and of the Council of 21 October 2009 establishing a framework for the setting of ecodesign requirements for energy-related products (Text with EEA relevance)

Waste Legislation
- **Waste Framework Directive** – Directive 2008/98/EC of the European Parliament and of the Council of 19 November 2008 on waste and repealing certain Directives
- **List of wastes** – 2014/955/EU: Commission Decision of 18 December 2014 amending Decision 2000/532/EC on the list of waste pursuant to Directive 2008/98/EC of the European Parliament and of the Council Directive 91/689/EEC on hazardous waste
- **Packaging Waste Directive** – European Parliament and Council directive 94/62/EC of 20 December 1994 on packaging and packaging waste

Immission protection

- **EU Action Plan: 'Towards Zero Pollution for Air, Water and Soil'** - Communication from the Commission to the European Parliament, the Council, the European Economic and Social Committee and the Committee of the Regions Empty COM(2021) 400 final
- **Directive on Ambient Air Quality** - Directive 2008/50/EC of the European Parliament and of the Council of 21 May 2008 on ambient air quality an cleaner air for Europe
- **Directive on industrial emissions** – Directive 2010/75/EU of the European Parliament and of the Council of 24 November 2010 on industrial emissions (integrated pollution prevention and control)
- **MCPD Directive** – Directive (EU) 2015/2193 of the European Parliament and of the Council of 25 November 2015 on the limitation of emissions of certain pollutants into the air from medium combustion plants
- **VOC emissions, paints and varnishes** – Directive 2004/42/CE of the European Parliament and of the Council of 21 April 2004 on the limitation of emissions of volatile organic compounds due to the use of organic solvents in certain paints and varnishes and vehicle refinishing products and amending Directive 1999/13/EC

- **Emissions Trading Directive** – Directive 2003/87/EC of the European Parliament and of the Council of 13 October 2003 establishing a scheme for greenhouse gas emission allowance trading within the Community and amending Council Directive 96/61/EC
- **PRTR Regulation** – Regulation (EC) No 166/2006 of the European Parliament and of the Council of 18 January 2006 concerning the establishment of a European Pollutant Release and Transfer Register and amending Council Directives 91/689/EEC and 96/61/EC
- **F-Gases Regulation** – Regulation (EU) No 517/2014 of the European Parliament and of the Council of 16 April 2014 on fluorinated greenhouse gases and repealing Regulation (EC) No 842/2006

Water Legislation

- **Water Framework Directive** – Directive 2000/60/EC of the European Parliament and of the Council of 23 October 2000 establishing a framework for Community action in the field of water policy

Hazardous Materials Transportation Law

- **Directive 2008/68/EC** of the European Parliament and of the Council of 24 September 2008 on the inland transport of dangerous goods
- **Council Directive 95/50/EC** of 6 October 1995 on uniform procedures for checks on the transport of dangerous goods by road
- **Regulation (EC) No 2099/2002** of the European Parliament and of the Council of 5 November 2002 establishing a Committee on Safe Seas and the Prevention of Pollution from Ships (COSS) and amending the Regulations on maritime safety and the prevention of pollution from ships
- **Directive 2010/35/EU** of the European Parliament and of the Council of 16 June 2010 on transportable pressure equipment and repealing Council Directives 76/767/EEC, 84/525/EEC, 84/526/EEC, 84/527/EEC and 1999/36/EC

International Treaties
GHS – Globally Harmonized System of Classification and Labelling of Chemicals
ADR – Agreement concerning the International Carriage of Dangerous Goods by Road
ADN – European Agreement concerning the International Carriage of Dangerous Goods by Inland Waterways
RID – The Regulation concerning the International Carriage of Dangerous Goods by Rail

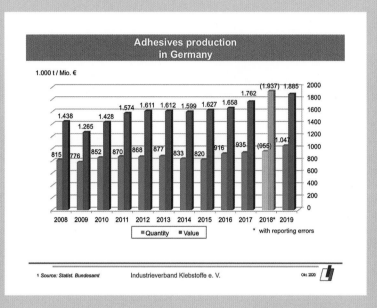

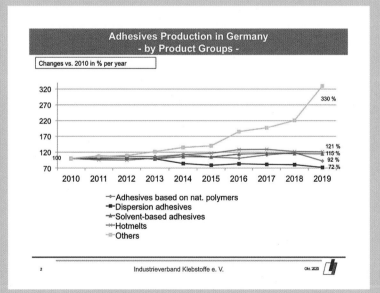

The German Adhesives Industry 2020
COVID-19-Pandemic leads global economy into deep recession

Geopolitical risks

➢ Interfered supply chains
➢ Further dept trends
➢ Escalation of trade disputes
➢ Brexit
➢ Unemployment

Historical decline in industry production

➢ Reduction of economic performance in all developed national economies
➢ Decreasing global IPX
➢ Slow recovery expected

Exchange Rates

➢ Development unpredictable, impacted by COVID-19-Pandemic

Raw Materials

➢ Increasing oil prices but still moderate level
➢ Despite Covid-19 effects still stable raw material situation

→ The German economy experienced severe recession. Many market segments suprisingly robust – further Recovery expected; service sector slower and rather long-term

Sources: IHS World Economic Service August 2020; DIW Berlin August 2020

Economy in recession – Dependend on the development of the COVID-19 pandemic only slow recovery expected

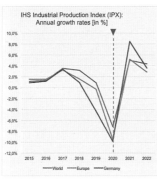

IHS Industrial Production Index (IPX): Annual growth rates [in %]

—World —Europe —Germany

➢ **Global economy declines in 2020**
Continuing demand decrease and supply chain disruptions; despite first signs of recovery still risks due to increasing infection numbers and lockdowns.

➢ **Strong decrease of the economy in Europe in 2020**, Italy and Spain particularly negatively impacted. Rising public deficit and debt are possible long-term problems.

➢ **Drastic economy decline in Germany in 2020**, as a result of lockdown measures. Uncertainty leads to restraints in purchasing investment goods and consumer goods.

Sources: IHS World Economic Service August 2020; DIW Berlin August 2020

The German Adhesives Industy
- Development of selected customer industries in Germany-

	Share in of total	2018	Prognosis 2019	Prognosis 2020
Manufacturing	**100**	**1,1**	**- 4,6**	**- 8,1**
Transportation	24,1	- 0,7	- 9,4	- 15,8
Food, Beverages and Tobacco	9,2	- 0,2	- 0,5	2,8
Papier / Printed Matters	3,0	- 1,0	- 3,4	- 3,1
Metals and Metal Products	12,7	0,9	- 2,8	- 6,5
Plant and Machinery	13,1	2,1	- 3,8	- 11,7
Electrical and Optical Equipment	10,8	1,6	- 9,0	- 11,9
Chemical Industry	7,3	- 2,1	- 3,9	- 2,9
Woodworking (withot Furniture)	1,3	2,0	- 1,6	- 3,2
Construction Industry	**--**	**0,2**	**3,4**	**- 1,3**

Source: IHS World Industry Service April 2020

Industrieverband Klebstoffe e. V.

Okt. 2020

5

The German Adhesives Industry
- Market Sentiment Survey -

Current Development
Sentiment in Response Area

- Market Situation
- Business Situation
- Sales Development

Source: IVK

Industrieverband Klebstoffe e. V.

Okt. 2020

6

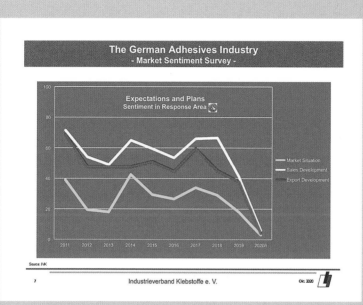

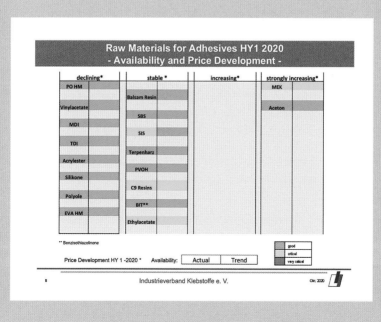

STANDARDS

Adhesives standards

The process of introducing national standards for adhesives began in Germany in the early 1950s. Subsequently a terminology standard and other standards for various fields (wood, shoe and flooring adhesives) were published. DIN standards issued during that period by the German Institute for Standardisation (DIN) still apply within Germany, unless they have been repealed or replaced by European standards. On 23 January 2013, DIN formed the new "Adhesives; Test Procedures and Requirements" committee, which functions as an umbrella committee for the DIN standards committees that deal with adhesives. The committee was created because most of the necessary standards are now available in the form of DIN EN standards, which means that the national standards committees are inactive. This consequence of this was that, in those cases where a revision was decided on at a European level, no active mirror committee and therefore no possibilities for participation were available. The formation of the "Adhesives; Test Procedures and Requirements" committee represented a milestone in the reorganisation of adhesives committees at DIN. It is now possible to spin off new subcommittees relatively quickly and therefore to react more rapidly to new requirements than was previously the case. Because of this and because of the integration of other DIN committees dealing with adhesives, the NMP (Standards Committee Materials Testing) sees itself as being well equipped for the future.

In 1961, the Comité Européen de Normalisation (European Committee for Standardization, CEN) was founded. Subsequently CEN established various technical committees (Technical Committee 52 "Safety of toys", Technical Committee 67 "Adhesives for tiles", Technical Committee 193 "Adhesives", Technical Committee 261 "Packaging" and Technical Committee 264 "Indoor air quality") to establish and revise standards in the field of adhesives. For the first time it was possible to develop a comprehensive, coherent set of standards that applied throughout Europe and that were specifically aimed at adhesives manufacturers and consumers. This also allowed adhesives manufacturers to represent their own economic and technical interests more effectively. Agreements which CEN signed with the International Organization for Standardization (ISO) meant that the global ISO standards were accepted as European standards and the two organisations went on to jointly draw up and issue additional standards.

European standards are generally published in the three official languages of the European Union – German, English and French – and are consistent with one another from a technical perspective. European standardisation supports trade across the EU by reducing trade barriers and is part of the framework of efforts to achieve technical harmonisation. European standardisation is supported by the German Adhesives Association. Unlike other standards, such as international ISO standards, European standards (EN) are binding at a European level. Both the European Court of Justice and national courts within the EU are obliged to base their rulings on European standards.

Specialist testing of adhesives and bonded joints is an important prerequisite for the success of the bonding process. The German Adhesives Association (IVK) has developed an online tool, which is available at www.klebstoffe.com, to provide users with support in this area.

The tool is a database that for the first time summarises all the relevant standards, directives and other important guidelines at a glance. The list currently consists of more than 800 individual documents. It includes not only the relevant German, European and international standards (DIN, EN, ISO), but also standards that were not specifically created for use in the field of bonding, but have become well established in this area.

All the documents are categorised on the basis of the type of adhesive, the application and the contents of the document. Practical search functions allow users to find the subject areas that are relevant to them. A brief description of the contents of each document helps users to determine whether it answers their questions or meets their specific testing requirements. If the document is available free of charge, there is a link to the original version. If payment is required, a link is provided to the website where the document can be purchased.

The standards database is kept constantly updated by IVK, in collaboration with DIN Software GmbH and the consultancy company KLEBTECHNIK Dr. Hartwig Lohse e.K. This ensures that users can always identify which testing methods for adhesives and bonded joints meet their needs.

The standards list is currently being redesigned and will soon be available on the IVK website with a new layout and a more convenient, product-based search function.

http://www.klebstoffe.com/start-normentabelle.html

SOURCES

- ▸ Raw Materials
- ▸ Adhesives by Types
- ▸ Sealants
- ▸ Adhesives by Key Market Segments
- ▸ Equipment
- ▸ Technical Consultancy
- ▸ Contract Manufacturing and Filling Services
- ▸ Research and Development

Raw Materials

Alberdingk Boley
ARLANXEO
Arakawa Europe
ASTORtec
Avebe Adhesives
BASF
BCD Chemie
Biesterfeld Spezialchemie
Bodo Möller
Brenntag
BYK
Chemische Fabrik Budenheim
CHT Germany
CnP Polymer
Coim
Collall
Collano
Covestro
CSC JÄKLECHEMIE
CTA GmbH
DKSH GmbH
Dunlop Tech GmbH
EMS-Chemie
Evonik
Gluetec Industrieklebstoffe
Gustav Grolman
Hansetack
IMCD
Jobachem
KANEKA Belgium
Keyser & Mackay
Krahn Chemie
LANXESS
Morchem
Nordmann, Rassmann
Möller Chemie
Münzing
Omya Hamburg GmbH
ORGANIK KIMYA
Poly-Chem
PolyU GmbH
Rain Carbon
Schill+Seilacher
SKZ – KFE

Synthomer Deutschland GmbH
Synthopol Chemie
Ter Hell
versalis S.p.A.
Wacker Chemie
Worlée-Chemie
WS INSEBO GmbH

Adhesives by Types

Hot-melt Adhesives
Adtracon
ALFA Klebstoffe AG
ARDEX
Artimelt
ASTORtec
Avery Dennison
BCD Chemie
Beardow Adams
Biesterfeld Spezialchemie
Bilgram Chemie
Bodo Möller
Bostik
BÜHNEN
BYLA
CHT Germany
Collano
CSC JÄKLECHEMIE
Dupont
Drei Bond
Eluid Adhesive
EMS-Chemie
EUKALIN
Evonik
Fenos AG
Follmann
GLUDAN
Gyso
H.B. Fuller
Fritz Häcker
Henkel
Jobachem
Jowat
Kleiberit
Kömmerling

L&L Products Europe
Morchem
Paramelt
Planatol
Poly-Chem
PRHO-CHEM
Rampf
Ruderer Klebtechnik
SABA Dinxperlo
Sika Automotive
Sika Deutschland
SKZ – KFE
Tremco illbruck
TSRC (Lux.) Corporation
Türmerleim
versalis S.p.A.
VITO Irmen
Weiss Chemie + Technik
Zelu

Reactive Adhesives
Adtracon
ARDEX
ASTORtec
BCD Chemie
Biesterfeld Spezialchemie
Bona
Bodo Möller
Bostik
BÜHNEN
BYLA
Chemetall
COIM Deutschland
Collano
CSC JÄKLECHEMIE
Cyberbond
DEKA
DELO
Drei Bond
Dupont
Dymax Europe
Fenos AG
fischerwerke
Gluetec Industrieklebstoffe
Gößl + Pfaff

H.B. Fuller
Henkel
Jobachem
Jowat
Kiesel Bauchemie
Kleiberit
Kömmerling
L&L Products Europe
Morchem
LORD
LOOP
LUGATO CHEMIE
merz+benteli
Minova CarboTech
Otto-Chemie
Panacol-Elosol
Paramelt
PCI
Planatol
Poly-clip System
PolyU GmbH
Rampf
Ramsauer GmbH
Ruderer Klebtechnik
SABA Dinxperlo
Schlüter
Schomburg
Sika Automotive
Sika Deutschland
SKZ – KFE
Sonderhoff
STAUF
Stockmeier
Synthopol Chemie
Tremco illbruck
Unitech
Uzin Tyro
Uzin Utz
Vinavil
Wakol
Weicon
Weiss Chemie + Technik
Wöllner
WS INSEBO GmbH
ZELU CHEMIE

Dispersion Adhesives
ALFA Klebstoffe AG
ASTORtec
ATP adhesive systems AG
Avery Dennison
BCD Chemie
Beardow Adams
Biesterfeld Spezialchemie
Bilgram Chemie
Bison International
Bodo Möller
Bona
Bostik
BÜHNEN
CHT Germany
Coim
Collall
Collano
CSC JÄKLECHEMIE
CTA GmbH
DEKA
ekp Coatings
Drei Bond
Eluid Adhesive
EUKALIN
Fenos AG
fischerwerke
Follmann
H.B. Fuller
GLUDAN
Grünig KG
Gyso
Fritz Häcker
Henkel
IMCD
Jobachem
Jowat
Kiesel Bauchemie
Klebstoffwerk COLLODIN
Kleiberit
Kömmerling
LORD
LUGATO CHEMIE
Morchem
Murexin
ORGANIK KIMYA

Paramelt
PCI
Planatol
Poly-clip System
PRHO-CHEM
Ramsauer GmbH
Renia-Gesellschaft
Ruderer Klebtechnik
Schlüter
Schomburg
Sika Automotive
SKZ - KFE
Sopro Bauchemie
STAUF
Synthopol Chemie
Tremco illbruck
Türmerleim
UHU
VITO Irmen
Wakol
Weiss Chemie & Technik
WS INSEBO GmbH
Wulff
ZELU CHEMIE

Vegetable Adhesives,
Dextrin and Starch Adhesives
BCD Chemie
Beardow Adams
Biesterfeld Spezialchemie
Bodo Möller
Collall
Distona AG
Eluid
EUKALIN
Grünig KG
H.B. Fuller
Henkel
Paramelt
Planatol
PRHO-CHEM
Ruderer Klebtechnik
SKZ - KFE
Türmerleim
Wöllner

Animal Glue
H.B. Fuller
Henkel
PRHO-CHEM

Solvent-based Adhesives
Adtracon
ASTORtec
Avery Dennison
BCD Chemie
Biesterfeld Spezialchemie
Bilgram Chemie
Bison International
Bodo Möller
Bona
Bostik
CHT Germany
COIM Deutschland
Collall
CSC JÄKLECHEMIE
CTA GmbH
DEKA
ekp Coatings
Distona AG
Fenos AG
Fermit
fischerwerke
Gluetec Industrieklebstoffe
Gyso
H.B. Fuller
IMCD
Jobachem
Jowat
Kiesel Bauchemie
Kleiberit
Kömmerling
LANXESS
LORD
Otto-Chemie
Paramelt
Planatol
Poly-Chem
Poly-clip System
Ramsauer GmbH
Renia-Gesellschaft
Ruderer Klebtechnik

SABA Dinxperlo
Sika Automotive
STAUF
Synthopol Chemie
Tremco illbruck
TSRC (Lux.) Corporation
UHU
versalis S.p.A.
VITO Irmen
Wakol
Weiss Chemie + Technik
ZELU CHEMIE

Pressure-Sensitive Adhesives
ALFA Klebstoffe AG
ASTORtec
ATP adhesive systems AG
Avery Dennison
BCD Chemie
Beardow Adams
Biesterfeld Spezialchemie
Bostik
BÜHNEN
Collano
CSC JÄKLECHEMIE
CTA GmbH
DEKA
Dymax Europe
Eluid Adhesive
EUKALIN
H.B. Fuller
GLUDAN
Fenos AG
Fritz Häcker
Henkel
IMCD
Jobachem
Kleiberit
L&L Products Europe
LANXESS
ORGANIK KIMYA
Paramelt
Planatol
Poly-Chem
PRHO-CHEM
Ruderer Klebtechnik

Sealants

ARDEX
Bodo Möller
Bison International
Bostik
Botament
CTA GmbH
Drei Bond
EMS-Chemie
Fermit
fischerwerke
Gluetec Industrieklebstoffe
Henkel
Jobachem
L&L Products Europe
merz+benteli
Murexin
ORGANIK KIMYA
OTTO-Chemie
Paramelt
PCI
Poly-clip System
PolyU GmbH
Rampf
Ramsauer GmbH
Ruderer Klebtechnik
Schomburg
SKZ - KFE
Sonderhoff
Stockmeier
Synthopol Chemie
Tremco illbruck
Unitech
UHU
WS INSEBO GmbH
Wulff

Adhesives by
Key Market Segments

Self Adhesive Tapes
Alberdingk Boley
Artimelt
ATP adhesive systems AG
Avebe Adhesives

Bodo Möller
BYK
certoplast Technische Klebebänder
CNP-Polymer
Coroplast
DKSH GmbH
Eluid Adhesive
Fritz Häcker
IMCD
LANXESS
Lohmann
Planatol
Schlüter
Synthopol Chemie
Tesa

Paper/Packaging
Adtracon
Alberdingk Boley
ALFA Klebstoffe AG
Arakawa Europe
Artimelt
ATP adhesive systems AG
Avebe Adhesives
BCD Chemie
Beardow Adams
Biesterfeld Spezialchemie
Bilgram Chemie
Bison International
Bodo Möller
Bostik
Brenntag
BÜHNEN
BYK
certoplast Technische Klebebänder
CNP-Polymer
COIM Deutschland
Collano
Coroplast
CSC JÄKLECHEMIE
CTA GmbH
DEKA
DKSH GmbH
ekp Coatings
Eluid Adhesive

EMS-Chemie
EUKALIN
Evonik
Fenos AG
Follmann
Gustav Grolman
Gyso
H.B. Fuller
GLUDAN
Grünig KG
Fritz Häcker
Hansetack
Henkel
IMCD
Jobachem
Jowat
LANXESS
Lohmann
Morchem
Möller Chemie
MÜNZING
Nordmann, Rassmann
Nynas
Omya Hamburg GmbH
ORGANIK KIMYA
Paramelt
Planatol
Poly-Chem
PRHO-CHEM
Ruderer Klebtechnik
Synthopol Chemie
tesa
TSRC (Lux.) Corporation
Türmerleim
UHU
versalis S.p.A.
Wakol
Weicon
Weiss Chemie + Technik
Wöllner

Bookbinding/
Graphics Industry
ALFA Klebstoffe AG
Arakawa Europe

ATP adhesive systems AG
BCD Chemie
Biesterfeld Spezialchemie
Bodo Möller
Brenntag
BÜHNEN
BYK
CNP -Polymer
Coim
Collall
CSC JÄKLECHEMIE
DKSH GmbH
Eluid Adhesive
EUKALIN
Evonik
Gluetec Industrieklebstoffe
Gustav Grolman
H. B. Fuller
Fritz Häcker
Hansetack
Henkel
IMCD
Jobachem
Jowat
LANXESS
Lohmann
Möller Chemie
MÜNZING
Nordmann, Rassmann
Omya Hamburg GmbH
ORGANIK KIMYA
Planatol
PRHO-CHEM
Sika Automotive
tesa
TSRC (Lux.) Corporation
Türmerleim
UHU
versalis S.p.A.
Vinavil

Wood/Furniture industry
Adtracon
ALFA Klebstoffe AG
Arakawa Europe

ATP adhesive systems AG
BCD Chemie
Biesterfeld Spezialchemie
Bilgram Chemie
Bison International
Bodo Möller
Bostik
Brenntag
BÜHNEN
BYK
BYLA
Chemische Fabrik Budenheim
CNP-Polymer
Collall
Collano
Coroplast
CSC JÄKLECHEMIE
CTA GmbH
Cyberbond
DEKA
DKSH GmbH
Eluid Adhesive
EMS-Chemie
Evonik
Fenos AG
fischerwerke
Follmann
Gluetec Industrieklebstoffe
Gößl + Pfaff
Gustav Grolman
Grünig KG
Gyso
Hansetack
H.B. Fuller
Henkel
Jobachem
Jowat
KANEKA Belgium
Kleiberit
Kömmerling
LANXESS
Lohmann
Minova CarboTech
Möller Chemie
Morchem

MÜNZING
Nordmann
Omya Hamburg GmbH
ORGANIK KIMYA
Otto-Chemie
Panacol-Elosol
PolyU GmbH
Rampf
Ramsauer GmbH
Ruderer Klebtechnik
SABA Dinxperlo
Sika Automotive
SKZ - KFE
STAUF
Stockmeier
Synthopol Chemie
tesa
Tremco illbruck
TSRC (Lux.) Corporation
Türmerleim
versalis S.p.A.
Vinavil
VITO Irmen
Wakol
Weicon
Weiss Chemie + Technik
Wöllner
WS INSEBO GmbH
ZELU CHEMIE

Building and Construction Industry
including Floors, Walls and Ceilings
ARDEX
artimelt
ASTORtec
ATP adhesive systems AG
BCD Chemie
Biesterfeld Spezialchemie
Bilgram Chemie
Bodo Möller
Bona
Bostik
Botament
Brenntag
BÜHNEN

BYLA
BYK
certoplast Technische Klebebänder
Chemische Fabrik Budenheim
CnP Polymer
Collano
Coroplast
CSC JÄKLECHEMIE
CTA GmbH
DEKA
DELO
DKSH GmbH
Dunlop Tech GmbH
Emerell
EMS-Chemie
Evonik
Fenos AG
Fermit
fischerwerke
Gluetec Industrieklebstoffe
Gößl + Pfaff
Gustav Grolman
Gyso
H. B. Fuller
GLUDAN
Gyso
Hansetack
Henkel
Jobachem
IMCD
Kiesel Bauchemie
Kleiberit
Kömmerling
Lohmann
LUGATO CHEMIE
Mapei
Minova CarboTech
Möller Chemie
Murexin
MÜNZING
Nordmann, Rassmann
ORGANIK KIMYA
Otto-Chemie
Paramelt
PCI

Planatol
Poly-Chem
Poly-clip System
PolyU GmbH
Rampf
Ramsauer GmbH
Schlüter
Schomburg
SCIGRIP Europe
Sika Automotive
Sika Deutschland
SKZ - KFE
Sopro Bauchemie
STAUF
Synthopol Chemie
tesa
Tremco illbruck
TSRC (Lux.) Corporation
Uzin Tyro
Uzin Utz
Vinavil
Wakol
Weicon
Weiss Chemie + Technik
Wöllner
WS INSEBO GmbH
Wulff

Car and Aircraft Industries
ALFA Klebstoffe AG
APM Technica
Arakawa Europe
ASTORtec
ATP adhesive systems AG
Beardow Adams
Bison International
Bodo Möller
Brenntag
BÜHNEN
BYLA
certoplast Technische Klebebänder
Chemetall
Chemische Fabrik Budenheim
CHT Germany
CNP-Polymer

Coroplast
CSC JÄKLECHEMIE
Cyberbond
DEKA
DELO
Drei Bond
Dunlop Tech GmbH
Dupont
Dymax Europe
Emerell
EMS-Chemie
Evonik
Fenos AG
Gluetec Industrieklebstoffe
Gößl + Pfaff
Gustav Grolman
H.B. Fuller
Hansetack
Henkel
Jobachem
Kleiberit
Kömmerling
L&L Products Europe
Lohmann
LORD
Möller Chemie
MÜNZING
Nordmann, Rassmann
Otto-Chemie
Panacol-Elosol
Planatol
Poly-clip System
Polytec
Rampf
Ramsauer GmbH
Ruderer Klebtechnik
SCIGRIP Europe
Sika Automotive
Sika Deutschland
Sonderhoff
Synthopol Chemie
Tremco illbruck
tesa
TSRC (Lux.) Corporation
Unitech

VITO Irmen
Wakol
Weicon
Weiss Chemie + Technik
ZELU CHEMIE

Electronics
APM Technica
ASTORtec
ATP adhesive systems AG
Bison International
Bodo Möller
Brenntag
BÜHNEN
BYLA
certoplast Technische Klebebänder
Chemetall
CHT Germany
Collano
Coroplast
CSC JÄKLECHEMIE
CTA GmbH
Cyberbond
DELO
DKSH GmbH
Drei Bond
Dymax Europe
Emerell
EMS-Chemie
Evonik
Gluetec Industrieklebstoffe
Gößl + Pfaff
Gustav Grolman
H.B. Fuller
Hansetack
Henkel
Jobachem
KANEKA Belgium
Kömmerling
L&L Products Europe
Lohmann
LORD
Möller Chemie
Morchem
MÜNZING

Nordmann, Rassmann
Otto-Chemie
Panacol-Elosol
Polytec
Rampf
Ruderer Klebtechnik
SCIGRIP Europe
Sika Automotive
Sika Deutschland
SKZ - KFE
tesa
Tremco illbruck
Unitech
UHU
Weicon
Weiss Chemie + Technik

Sanitary Industry
APM Technica
Arakawa Europe
Bilgram Chemie
CSC JÄKLECHEMIE
H.B. Fuller
GLUDAN
Gustav Grolman
Henkel
Jowat
Kömmerling
LANXESS
Lohmann
Nordmann, Rassmann
Prho-Chem
Sika Automotive
Türmerleim
Vito Irmen

Assembly, General Industry
ASTORtec
BCD Chemie
Biesterfeld Spezialchemie
Bodo Möller
BÜHNEN
BYLA
certoplast Technische Klebebänder
Chemetall

CHT Germany
Coroplast
CSC JÄKLECHEMIE
Cyberbond
DEKA
DELO
Drei Bond
Dupont
Gößl + Pfaff
Henkel
KANEKA Belgium
Kleiberit
Kömmerling
L&L Products Europe
Lohmann
Otto-Chemie
Panacol-Elosol
Paramelt
Renia-Gesellschaft
Ruderer Klebtechnik
SABA Dinxperlo
Schomburg
SCIGRIP Europe
SKZ - KFE
Synthopol Chemie
tesa
Weicon

Textile Industry
Adtracon
ASTORtec
ATP adhesive systems AG
BCD Chemie
Biesterfeld Spezialchemie
Bodo Möller
Bostik
Brenntag
BÜHNEN
Chemische Fabrik Budenheim
CHT Germany
CNP-Polymer
Collano
CSC JÄKLECHEMIE
DEKA
Emerell

EMS-Chemie
EUKALIN
Evonik
Gluetec Industrieklebstoffe
Gustav Grolman
Hansetack
H.B. Fuller
Henkel
Jobachem
Jowat
Kleiberit
LANXESS
Möller Chemie
Morchem
MÜNZING
Nordmann, Rassmann
Omya Hamburg GmbH
SABA Dinxperlo
Sika Automotive
Synthopol Chemie
tesa
Vito Irmen
Wakol
Wulff
Zelu

Gyso
Hansetack
H.B. Fuller
Henkel
IMCD
Jobachem
Jowat
KANEKA Belgium
LANXESS
Möller Chemie
MÜNZING
Nordmann, Rassmann
Nynas
ORGANIK KIMYA
Paramelt
Planatol
PRHO-CHEM
Stauf
Sika Automotive
SKZ - KFE
Synthopol Chemie
TSRC (Lux.) Corporation
Türmerleim
versalis S.p.A.
Vito Irmen

Self Adhesive Tapes, Labels
artimelt
Arakawa Europe
ASTORtec
ATP adhesive systems AG
Avery Dennison
BCD Chemie
Biesterfeld Spezialchemie
Bodo Möller
Bostik
Brenntag
CNP-Polymer
Coim
Collano
EMS-Chemie
EUKALIN
Fenos AG
Gluetec Industrieklebstoffe
Gustav Grolman

Household, Hobby,
Offices, Stationery
Arakawa Europe
Bodo Möller
Bison International
certoplast Technische Klebebänder
CNP-Polymer
Collall
Coroplast
CSC JÄKLECHEMIE
CTA GmbH
Cyberbond
EMS-Chemie
Fenos AG
Fermit
fischerwerke
GLUDAN
Gustav Grolman
Gyso

Hansetack
Henkel
Jobachem
KANEKA Belgium
LUGATO Chemie
Möller Chemie
Nordmann, Rassmann
Nynas
Omya Hamburg GmbH
Panacol-Elosol
Rampf
Ramsauer GmbH
Renia-Gesellschaft
SCIGRIP Europe
tesa
Tremco illbruck
TSRC (Lux.) Corporation
UHU
versalis S.p.A.
Weicon
Weiss Chemie + Technik
WS INSEBO GmbH

Footwear & Leather Industry
Adtracon
BÜHNEN
Cyberbond
H.B. Fuller
Henkel
Kömmerling
Renia-Gesellschaft
Ruderer Klebtechnik
Sika Automotive
Wakol
Zelu

Equipment
for Adhesive Handling, Mixing, Dosing
and Application
Baumer hhs
bdtronic
Beinlich Pumpen
BÜHNEN

Drei Bond
Hardo
Hilger u. Kern
Hönle
H&H Maschinenbau
Innotech
IST Metz
Nordson
Plasmatreat
Reinhardt-Technik
Reka Klebetechnik
Robatech
Rocholl
Scheugenpflug
Sulzer
Unitechnologies SA – mta
ViscoTec Pumpen- u. Dosiertechnik
VSE Volumentechnik
Walther

Technical Consultancy
ChemQuest Europe INC.
Hinterwaldner Consulting
Klebtechnik Dr. Hartwig Lohse

Contract Manufacturing and Filling Services
LOOP

Research and Development
IFAM
BFH
SKZ – KFE
ZHAW